GUAR IN INDIA

The Editors

Dr. D. Kumar is currently working as Principal Scientist (Plant Breeding) and is holding the charge of Project Coordinator, All India Coordinated Research Project on Arid Legumes, at Central Arid Zone Research Institute, Jodhpur. He has edited 9 books and has published more than 120 full-length research papers in different journals of national and international reputes. He has also written more than 20 chapters for different books and Proceedings of the Symposia. His main feld of specialization is Breeding for Abiotic Stress Tolerance with special reference to drought and salinity in arid pulses and rapeseed-mustard.

Dr. N.B. Singh is currently working as Assistant Director General (Oilseeds & Pulses) at the ICAR, New Delhi. Dr. Singh is Ph.D. in Plant Breeding from the reputed University of Minnesota, USA. He has served Rajendra Agricultural University, Pusa (Bihar) as Professor of Plant Breeding for more than one decade. As distinguished Professor, he has served International Crops Research Institute for Semi-Arid Tropics (ICRISAT), Hyderabad as visiting scientist for more than three years. He has contributed a number of research papers in the area of Pulse Crops Improvement in national and international journals of repute. He has long experience of working at ICAR, New Delhi as Assistant Director General of Pulses and Oilseeds. He is well recognized for his research contributions in both pulses and oilseeds.

GUAR IN INDIA

Editors
D. Kumar
N.B. Singh

SCIENTIFIC PUBLISHERS (INDIA)
P.O. BOX 91 JODHPUR

Published on behalf of
INDIAN ARID LEGUMES SOCIETY

by

Scientific Publishers (India)
5-A, New Pali Road, P.O. Box 91,
Jodhpur - 342 001
Tel.: +91-291-433323
Fax : +91-291-512580
E-mail : scienti@sancharnet.in
artrans@satyam.net.in
http://www.scientificpub.com

ISBN: 81-7233-310-2

Lasertypeset : Rajesh Ojha
Printed in India

भाकृअनुप
ICAR

E-mail : mrai@icar.delhi.nic.in
Fax : 91-11-3382545
Telex : 031-62249 ICAR.IN
Telegram : AGRISEC
Phone : Office : 3382545
3388991/567
Resi. : 5852453

DR. MANGALA RAI
Deputy Director General
(Crop Science)

INDIAN COUNCIL OF AGRICULTURAL RESEARCH
Krishi Bhawan, Dr. Rajendra Prasad Road, New Delhi-110 001

FOREWORD

Guar is a drought-hardy multi-purpose crop, predominantly cultivated as a rainfed crop in arid and semi-arid regions in India. In Rajasthan, the principal guar-growing state, its acreage and productivity is, however, greatly fluctuating owing to uncertainty in distribution pattern of rainfall. Consistency and stability in production is thus the need of the time as guar forms the major source of livelihood for marginal farmers in Rajasthan. Efforts in this direction alone can help in harnessing the crop's tremendous potential, particularly as an industrial crop, so as to significantly increase its potential as foreign exchange earner, considering the use of guar gum in more than 200 products, including its use in oil and textile industries. The multiple industrial uses of the crop have amply indicated that the current foreign exchange earnings to the tune of about Rs. 800.00 crores to the country is far below the scope in this crop of immense promise.

Additionally, the crop's major consumption as food and important fodder source, besides its excellent qualities as soil improving legume further underscores the need for concerted efforts in its promotion in important guar growing areas to help move towards sustainable farming fragile agro-ecologies of arid and semi-arid areas. Compilation of research results pertaining to guar production and processing technologies is thus a laudable step. I, therefore, congratulate Drs. D. Kumar and N.B. Singh for their sincere efforts, and hope that all concerned with guar research and development would find the book "**Guar in India**" informative and useful.

Date : 15.3.2002

(MANGLA RAI)

PREFACE

Clusterbean (*Cyamopsis tetragonoloba* (L.) Taub.) is more commonly known as "guar" at the national and international levels. Guar has unique adaptatation and potential to yield under adverse, harsher and uncongenial soil and atmospheric situations, when majority of the crops may fail even to produce biomass. Specifically adapted to poor inputs and less care, guar is the most preferred priority crop of the arid farmers particularly, under weak, delayed and highly erratic monsoonic rains. Being an important component of cropping system in arid and semi-arid regions, it is a source of multi uses - green vegetables, fodder, feed, soil enrichment through nitrogen fixation, also a good source of Ca and Fe and vitamin A. Guar has, however, got great recognition due to presence of water soluble natural polymer galactomannan gum in protein free endosperm portion of the seeds. The galactomannan gum content due to its specific qualities as thickner, binder and stabilizer has verstalile uses in a number of industries. The export of guar gum is a big catch for the farmers and the industrialists as well as the same is exported to the Western countries and help earning foreign exchange. This has raised the status of guar from a neglected and marginal crop to an industrial crop of international recognition. There are more than 80 small and big guar gum industries in India, exporting guar split, powder or gum. This crop is, therefore, helping in employment generation also.

Inspite of having multi-uses with great recognition, guar is also known for greater fluctuations in area leading to unsustained productions over the years. Rajasthan is the principal guar growing state occupying 78% country's average. However, productivity of Rajasthan is lowest (233.7 kg ha^{-1}) than the National (322 kg ha^{-1}), Punjab (922 kg ha^{-1}), Haryana (740 kg ha^{-1}) and Gujarat's (668 kg ha^{-1}) productivity levels. Thus, attempts on balancing between higher areas with poor productivity in Rajasthan vis-a-vis low area and higher productivity in other states is the main focal strategy that would help sustainability in guar production. Besides, dissemination of convenient and economically feasible technologies are desired which can help increased productivity of guar in India. It would bridge the gap between realizable and realized yield potential of this arid legume in India.

In view of above, therefore, it would be quite pertinent if research results which are scattered and beyond the reach of real' users are compiled and composed in the form of a publication so as to make it available to the scientists, administrators, planners, industrialists and the other end users. It has, therefore, been tried to compile the research information on various aspects of improvement, production, protections, physiology, genetic resources, quality and industrial aspects etc. and brough out in the form of the book entitled 'Guar in India'.

The attempt on such classified information concerning guar are limited. It is possibly the first attempt at the National level that research results of scholary, practical and industrial relevance have been compressed and brought out. It is hoped that, this publication being published by newly formed society 'Arid Legumes Society of India" would help increase guar production in India through horizontal and vertical expansion. The book will provide a base for improving the quality and quantity of guar gum for export purpose and contributing significantly for the foreign exchequer.

D. Kumar

ACKNOWLEDGEMENTS

It gives me immense pleasure in acknowledging the help and inspiration exercised time and again by Dr. M. Rai, DDG (CS), ICAR, New Delhi. Dr. Rai has been great instrumental and the source of inspiration to me in the affairs of AICRP on Arid Legumes in general, and the compilation and publication of this book in a particular. Dr. Pratap Narayan, Director, CAZRI, Jodhpur for extending required facilities in bringing out this publication is also profoundly thanked.

Personal thanks are extended in favour of Dr. N.B. Singh ADG (O & P), ICAR headquarters, New Delhi, for timely help made available and inspired me at the needed hours.

The publication of this book would not have been possible without the contributions made by the fellow scientists engaged in guar research at the AICRP Centres. I particularly, thank Drs. J.V. Singh, R.N. Arora, B.D. Yadav, R.K. Kashyap and U.N. Joshi from CCS, HAU, Hisar, I am equally grateful to Dr. Virendra Kumar from RRS, CCS, HAU, Bawal. Thanks are also spared in favour of Mr. N.S. Yadav and R.L. Ray from Hindustan Gum and Chemicals Ltd., Bhiwani for writing a chapter on industrial Aspects. Dr. U.P. Singh and Dr. Dinesh K.Agarwal from IGFRI, Jhansi and Dr. S.Lal from Kanpur are also highly thanked for their valuable contributions made in due course.

I extend my sense of gratitudes in favour of Drs. N.K. Dwivedi and Neelam Bhatnagar from NBPGR, Regional Research Station, Jodhpur for valuable contribution on Genetic Resources of Clusterbean.

The help rendered and extended in contribution of respective chapters for the present book by my fellow colleagues from this Institute, is also greatly sensed. For instance, the names of Drs. A.V. Rao, B.K. Garg, S. Lodha, Uday Burman, Anurag Saxena and Mrs. Ritu Mawar are thankfully mentioned.

The acknowledgement is credited in favour of Manju Singh, Neetu Rathore and Seema Guar who have put up their maximum best in giving present shape to the book. It was their sincere and timely efforts that really led to bring out this publication. My thanks are due to Mr. P.S. Bhati for providing needed help. I share my words of thanks to Shri D.K. Nainwal also, who has been helping in typing work in the project and the contents of this book also.

Jodhpur
Dated : 10th May, 2002

D. Kumar

CONTRIBUTORS

1. Dr. D. Kumar
   Principal Scientist (Plant Breeding),
   CAZRI, Jodhpur - 342 003, Rajasthan.
2. Dr. N.B. Singh
   Assistant Director General (O & P),
   ICAR, Krishi Bhawan, New Delhi.
3. Dr. J.V. Singh
   Sr. Guar Breeder, Forage Section,
   CCS, HAU, Hisar - 125 004, Haryana
4. Dr. R.N. Arora
   Plant Breeder, Forage Section,
   CCS, HAU, Hisar - 125 004, Haryana.
5. Dr. B.D. Yadav
   Sr. Agronomist, Forage Section,
   CCS, HAU, Hisar - 125 004, Haryana.
6. Dr. Virendra Kumar
   Sr. Agronomist, RRS,
   CCS, HAU, Bawal, Haryana - 123 501
7. Dr. Anurag Saxena
   Sr. Scientist (Agronomy),
   CAZRI, Jodhpur - 342 003, Rajasthan.
8. Dr. B.K. Garg
   Principal Scientist (Plant Physiology),
   CAZRI, Jodhpur - 342 003, Rajasthan
9. Dr. Uday Burman
   Sr. Scientist (Plant Physiology),
   CAZRI, Jodhpur - 342 003, Rajasthan.
10. Dr. U.P. Singh
    Principal Scientist (Economic Botany),
    IGFRI, Jhansi 284 003, U.P.
11. Dr. Dinesh K. Agarwal
    Sr. Scientist (Economic Botany),
    IGFRI, Jhani - 284 003, U.P.

12. Dr. N.K. Dwivedi
Principal Scientist (Economic Botany), NBPGR,
RRS, CAZRI Campus, Jodhpur - 342 003, Rajasthan.

13. Dr. (Mrs.) Neelam Bhatnagar
Sr. Scientist (Economic Botany), NBPGR,
RRS, CAZRI Campus, Jodhpur - 342 003, Rajasthan.

14. Dr. A.V. Rao
Principal Scientist (Microbiology),
CAZRI, Jodhpur - 342 003, Rajasthan.

15. Dr. Satish Lodha
Principal Scientist (Plant Pathology),
CAZRI, Jodhpur - 342 003, Rajasthan.

16. Dr. Ritu Mawar
Sr. Research Fellow, CAZRI,
Jodhpur - 342 003, Rajasthan

17. Dr. R.K. Arora
Sr. Entomologist, SKUA & T, RHRS,
Udheywalla, Jammu Tawi - 180 002, J & K.

18. Dr. R.K. Kashyap
Sr. Entomologist, Department of Entomology,
CCS, HAU, Hisar - 125 004, Haryana.

19. Dr. U.N. Joshi
Sr. Biochemist, Dept. of Plant Breeding,
CCS, HAU, Hisar - 125 004, Haryana.

20. Mr. N.S. Yadav
Hindustan Gum and Chemicals Ltd.,
Birla Colony, Bhiwani - 127 021, Haryana.

21. Mr. R.L. Ray
Hindustan Gums and Chemicals Ltds.,
Birla Colony, Bhiwani - 127 021, Haryana.

22. Dr. S. Lal
Director (Retd.),
Indian Institute of Pulses Research, Kalyanpur,
Kanpur - 208 024, U.P.

CONTENTS

CHAPTER - 1

GUAR - AN INTRODUCTION

D. KUMAR and N.B. SINGH

The word guar [*Cyamopsis tetragonoloba* (L.) Taub.] represents its derivation from the Sanskrit word **"GAU AAHAR"** which means cow fodder or otherwise fodder of the livestock. The crop is also known by local names as Khutti, Dararietic, Guari etc. Basically, guar is a drought hardy, deep-rooted, summer annual legume. The crop is mainly grown in the dry habitats of Rajasthan, Haryana, Gujarat and Punjab and to limited extent in Uttar Pradesh and Madhya Pradesh. In addition to its major cultivation in India, the crop is also grown as a cash crop but to limited extent in other parts of the world, like, Australia (Anonymus, 1911), Brazil (Costa, 1950) and South Africa (Doidge, 1952). Guar was introduced from India to Oklohama plains of North Texas in 1903 (Hymowitz and Matlock, 1963) where, it has been cultivated quite successfully. The crop is known for exceptionally high adaptation towards poor and erratic rains, low inputs and less care, soil enrichment properties, multi uses in cropping systems, in industrial use in many ways and other social and dietry uses. These qualities have made it the most favoured crop of marginal farmers in the arid areas.

Acreage and Adaptation

Guar is grown in almost 23.30 lakh hectares area in India. The maximum contribution of states in respect of area is shared by Rajasthan (18.18 lakh ha), followed by Gujarat (2.27 lakh ha), Haryana (1.27 lakh ha) and Punjab (0.14 lakh ha). In other words, Rajasthan alone comprises almost 78.04% area, whereas, other states show quite negligible contribution to the tune of 9.74, 5.54 and 0.60%, respectively. However, 78.0% area of guar in Rajasthan alone, contributes almost 81.1% production to the national basket of guar.

The productivity (long term basis) however, shows just reverse trend wherein, productivity of guar in Rajasthan is lowest (233.7 kg ha^{-1}) against the national productivity (322.7 kg ha^{-1}). Productivity is, however, maximum (922.0 kg ha^{-1}) in Punjab, followed by Haryana (740.0 kg ha^{-1}), and Gujarat (668.8 kg ha^{-1}) states. Thus, guar yield at the national level could be increased by doubling the productivity in Rajasthan and increasing the area in other states like Haryana, Gujarat and Punjab, where productivity is already higher. Following table gives comparative figures on area, production and productivity for India and main guar growing states (Table 1).

Table 1. Area, production and productivity of guar in India and important states

India and States	Area (lakh ha)	Production (lakh ton)	Productivity* (kg ha^{-1})
India	23.30	11.98	428.0
Rajasthan	19.85	7.33	370.0
Gujarat	2.27	1.15	668.8
Haryana	1.20	1.08	740.0
Punjab	0.14	0.0065	922.0

* Long term basis
Figures pertaining to 1997-98

Botany

Guar belongs to the family leguminaceae and sub-family paplionaceae. It is a robust annual with long tap root and well developed laterals. It is generally 50 to 100 cm tall and bears 4-10 branches (branched types), however, unbranched type varieties having main stem only which is heavily clustered with pods. Guar plant may bear on an average basis 30-90 pods per plant. The leaves of this arid legume are alternate trifoliate, which are born on long petioles. The stem is tall and slender. The flowers are born in axillary raceme on long petioles. The flowers are small, typically papilionaceous. The five unequal linear teeth type sepals form the calyx while, papilioinaceous corolla has orbicular standard oblong with petals and long broad keel petals. The stamens are arranged in diadalphous condition. Style is short, while stigma is head shaped (Sindhu *et. al.,* 1982). The pods are oblong, 5-12 cm in length, normally containing 5-12 oval or cube shaped seeds of variable shape (Poats, 1961). The crop due to cleistogamous nature is considered strictly self pollinated, however, out crossing to the tune of 0.5 to 7.9% has also been reported (Saini *et. al.,* 1981).

Uses

Being a leguminous annual it's primary use is in soil health enrichment through atmospheric nitrogen fixation, on an average basis it has been estimated that it may fix nearly 30 kg N ha^{-1}. Furthermore, due to complete sheding of leaves up to maturity, the organic carbon may be added to the soil. From the ancient, guar is being used as a source of food, fodder and feed. Young pods are a source of delicious vegetable from late summer to the mid of rainy season. The pods are sweet and can often be cooked as those of *Phaseolus* bean pods. Guar seed meal a byproduct of guar gum industries consisting of seed coat and germ cell material, is a good source of protein. The same contains about 40-42% crude protein. It is principally used as a feed for livestock and poultry. Main and the most challenging use of guar is in various industries. The natural polysaccharide water-soluble polymer found in the endosperm, as galactomannan gum, is the chief product used in many industries. The use of guar gum can be envisaged in the system where water is an important factor. A number of foodstuff compositions require the use of colloids to modify the behaviour of water in relation to other compounds.

Guar gum acts as a binder, thickener or stabilizer for the following industrial products :

Food industries : Frozen foods, baked foods, processed cheese, dairy products, dressings and sauces, instant mixings, meat products, canned meats, beverages and pharmaceutical products.

Other industries : Textile thickness, paper industry, mining industry, oil well drilling, explosive industry, tobacco industry, fire fighting, industrial water treatment, agriculture etc.

Opportunities

Guar provides unique opportunities to the growers, consumers, industrialists, and environmentalists and also in employment generation. This is due to its ample usable properties. The same can be outlined below:

1. Being tolerant to drought, heat and showing medium tolerance to salinity, guar provides adequate opportunity to utilize such adversities in realization of production. In other words, under acute soil moisture deficit situations, encountered with highly erratic rains, nutritionally deficient soils having moderate salinity, this is the crop to be reckoned with. Guar appears as the immediate choice among large number of crops, in augmenting production and therefore, help comba- ting desertification and degradation of soil environment.
2. Guar has high transpiring ability leading to the cooling of canopy and increasing the soil moisture extraction from deeper layers.
3. Meristematic type nodules found in guar, provide greater recovery potential to this crop at the vegetative phase.
4. Guar pods are a very good source of vitamin A, calcium, iron, phosphorus and ascorbic acid, therefore, adding to it's nutritive values (Table 2).
5. Being a good source of vegetables, this crop provides good opportunity for kitchen and market gardening. On an, average basis, guar crop can give 37 (Pusa Sadabahar) to 160 q ha^{-1} (Sharad bahar) green pods in 50 to 90 days under irrigated conditions.
6. It provides sizeable number of employment in arid areas through guar gum industries swelling to almost 85 in Northern Western part of the country. It also generates employment through world trade comprising to the order of 1,20,000 ton annually, of which, India contributes nearly 75%.
7. Galactomannan gum content obtained from protein free splits, is a big catch and source of foreign exchange. For instance guar gum to the tune of 110,000 metric ton costing Rs. 814 crores was exported to Western countries in 1999-2000. Thus, growing of guar is good opportunity for the farmers and the industrialists as well. Guar is, therefore, a national opportunity for resource constraint persons depending for their livelihood on marginal lands.

Table 2. Nutritional composition of green guar pods

Constituent (per 100 g edible portion)	Content
Energy (kcl)	16
Moisture (g)	81
Protein (g)	3.2
Fat (g)	1.4
Carbohydrate (g)	10.8
Vitamin A (IU)	65.3
Thiamine (mg)	0.09
Riboflavin (mg)	0.03
Niacin (mg)	0.6
Ascorbic acid (mg)	49.0
Calcium (mg)	57.0
Phosphorus (mg)	57.0
Iron (mg)	4.5

Source : Roy and Chakarborti, 1993.

Challenges

Guar is known for versatile uses in various ways of livelihood and it provides good opportunities also in national agriculture sector through export potential. However, it has been facing a number of challenges also which are affecting it's area, production and therefore, shattering quantum of export also. Thus, it is desired to analyse such challenges preciously. Some of the critical challenges are outlined below :

1. Technical

Technically, the crop suffers serious bottlenecks inspite of being investigated at various levels in integrated manner for at least past two decades. Main technical challenges are as follows:

(a) Poor and unstable productivity : Guar productivity being influenced by various components has by and large been low and fluctuating over the years. In major guar growing State of Rajasthan for instance, productivity has been varying from 87 (1999-2000) to as high as 453 kg ha^{-1} (1990-91) against long term average productivity of 370 kg ha^{-1}. In Rajasthan, guar is facing low and fluctuating productivity whereas, in the States of Gujarat, Haryana and Punjab, crop is facing low acreage (Tabe 3). Thus, sustained productivity of guar in Rajasthan, may lead to furthering national productivity.

Table 3. Area, production and productivity of guar and rainfall in Rajasthan during the past 11 years

Year	Area (lakh ha)	Production (lakh ton)	Productivity (kg ha^{-1})	Rainfall (mm)
1990-91	20.89	9.46	453	437.93
1991-92	15.58	2.04	131	235.86
1992-93	18.22	5.83	320	449.57
1993-94	18.97	2.87	151	368.91
1994-95	19.59	7.08	361	428.02
1995-96	17.74	2.74	155	473.95
1996-97	18.18	7.39	407	551.43
1997-98	19.85	7.35	370	503.70
1998-99	16.11	3.19	198	505.89
1999-00	26.48	23.15	87	245.62
2000-01	30.56	4.81	157	238.56

(b) No variety for summer sowing : In irrigated areas, particularly in IGNP areas (Rajasthan), cultivation of guar is increasing in summer season, industry people are also showing great interest in this regard. However, a photo and thermo-insensitive variety, suited to summer season and responding to inputs, is not available, which is causing serious obstacle in its enhanced production during the off season.

(c) Susceptibility to diseases : Alternaria leaf spot, Bacterial blight, Anthracnose, root rot and wilts considerably damage this crop under the favourable influence of environment. However, in spite of a great surge in screening of varieties against these diseases, still there is no fool proof variety tolerant to more than one diseases at a time and place.

(d) Longer maturity : Due to longer maturity (85-110 days), guar varieties suffer from terminal drought under rainfed conditions, early varieties (<80 days) having drought escape mechanism with high yield are still not at all available. This is perhaps the biggest barrier for stabilization of yields of this crop under rainfed condition.

2. Socio-economic

Guar occupies marginal status and is considered second priority of the arid farmers. First priority for instance, being pearl millet for utilization of sufficient rains received during first week of July. Farmers prefer to plant guar at the late onset of rains and choose the soils having poor fertility and poor texture. Thus, there is need to raise the status of the crop from marginal to main crop and ensuring its planting at the early onset of monsoonic rains.

3. Support price

High production resulting from increased area during 1999-2000 may have been due to high market price during 1998-99 (touching Rs. 2500-2800 q^{-1}), whereas, during the current season prices have crashed down to just Rs. 800 q^{-1} in concomitant with high production. Thus, farmers are at a loss and there are no buyers. Support prices if provided for this crops, as available for other crops like pearl millet in arid zone will provide great incentive to the farmers.

4. Nutrition and Quality

Guar protein is characterized with low contents of sulphur containing amino acids (Methionine and Cystine). The combined percentage of S containing amino acids in guar is found only 1.11 compared to 4.00 mole percentage found in soybean. Thus, a genotype having higher proportion of soluble to insoluble protein fraction is desired for making it more acceptable at the nutritional level.

Biggest challenge guar crop now-a-days facing, is the availability of alternate sources of galactomannan gum or the hydrocolloids. For instance, *Sesbania bispinosa*, *Sesbania sesban*, *Cassia accidentails*, *Cassia sophera* and *Cassia tora* having abundant availability and comparative gum quality are causing serious threat for the cultivation of *C. tetragonoloba*. For instance, endosperm content (the location of gum) is higher in other legumes ranging from 33.0 (*C. tora*) to 37% (*S. bispinosa*) against 32.5% for guar. Thus, it may lead to high *per se* gum content in these species (Table 4). Higher viscosity desired for export purpose was also found considerably more in *S. seaban* (3160) and *S. bispinosa* (3800) than in *C. tetragonoloba* (2800 cps). Challenges currently being faced by guar crop due to the alternate sources of gum from non-traditional and uncommon legumes, can be thwarted by enhancing the productivity of guar grain yield and also enhancing guar gum content *per se*. For that ample opportunities exist.

Table 4. Comparative carbohydrate composition (%) and rheological attributes of various galactomannan gum containing legume seeds

Gum containing legumes	Endosperm (%)	Galacto-mannan (%)	Galactose: Mannose ratio (G/M)	Viscosity[a] $(n)^2$ cps	Rheological attributes		
					Consistency index value 'K'	Flow behaviour index, value 'n'	Rheological behaviour
Guar (*Cyamopsis tetragonaloba*)	32.5 ± 2.0	31.2 ± 2.0	1 : 20	2800.0	4.26	0.47	Pseudoplastic
Sesbanta bispinosa	37.5 ± 1.5	34.5 ± 1.5	1 : 2.20	3800.0	3.40	0.46	Pseudoplastic
Sesbania sesban	34.5 ± 1.5	31.8 ± 0.5	1 : 1.80	3160.0	2.98	0.45	Pseudoplastic
Cassla tora	33.0 ± 1.5	30.5 ± 0.5	1 : 1.50	1880.0	1.72	0.78	Pseudoplastic
Cassia sophera	36.0 ± 2.5	32.5 ± 1.25	1 : 2.30	1900.0	1.60	0.68	Pseudoplastic
Cassia accidentalis	35.2 ± 2.5	31.5 ± 1.0	1 : 2.60	1650.0	1.21	0.66	Pseudoplastic

a. The value represents at 1.0 w/v gum solution, at shear rate 4.51 s^{-1}, and temperature 30 ± 1.5°C.

Moreover, in the international market India's nearest and immediate competitor for gum export is Pakistan whose, gum has by and large been considered of better quality than ours.

Summary

Guar a principal arid legume, is known for versatile uses. Fluctuations in area and productivity of this crop in Rajasthan, is a matter of serious concern. For instance, area of guar varied from 15.58 (1991-92) to 30.56 lakh ha (2000-01) and productivity from 87.0 (1999-2000) to as high as 407.0 kg ha^{-1} (1996-97). For acceleration in the production of guar its productivity has to be, therefore, sustained in Rajasthan. It is required to keep pace in export of galactomannan gum to Western countries, so as to have considerable earning in foreign exchange. For instance, guar gum to the tune of Rs. 814 crore was exported during 1999-2000. Inspite of many uses, and opportunities, this crop is also facing serious challenges of varied nature. However, the most serious challenge is the availability of alternate source of gum content of almost similar quality. For instance, *Sesbania bispinosa*, *Sesbania sesban*, *Cassia accidentalis*, *Cassia tora* and *Cassia sophera* legume species have gum content of higher extent with high viscosity. For instance, *S. bispinosa* comprises 34.5% gum content with 3800 cps viscosity vis-a-vis 32.5% and 2800 cps for guar, respectively. Thus, in view to face this challenge, it is desired to increase guar gum content *per se* to the tune of 35.0% and productivity up to 600 kg ha^{-1}.

REFERENCES

Anonymous, 1911. *Guar*, Agric. Gazette, New Southwails, 22: 1000.

Costa, A.S. 1950. *Beta pettalaris*, a test plant for tobacco white necrosis virus. *Bragantia* 10: 275-276.

Doidge, E.M. 1952. *South African Fungi*. Pretoria National Herbm.

Hymowitz, T. and Matlock, R.S. 1963. Guar in the United States, *Okla Agril. Exptl. Stn. Bull*, B611: 3-14.

Poats, J.J. 1961. Guar a summer row crop for Southwest, *Econ. Bot.* 14: 241.

Roy, S.K., Chakarborti, A.K. 1993. Vegetables for temperate climate. *In: Encylopaedia of Food Science, Food Technology and Nutrition* (Ed. R. Ma Crae, R.K. Robinson and M.J. Sadler), Academic Press London, pp. 4715.

Saini, M.L., Arora, R.N. and Paroda, R.S. 1981. Morphology of three species of genus *Cyamopsis*. *Guar Newsl.* 2: 7-11.

Sindhu, A.S., Pandita, M.L., Arora, S.K. and Vashistha, R.N. 1982. Studies on genetic variability and correlations in clusterbean (*Cyamopsis tetragonoloba* (L.) Taub), *Haryana Agric. Univ. J. Res.* 12: 225-230.

CHAPTER - 2

GENETICS AND BREEDING

J. V. SINGH and R. N. ARORA

Clusterbean [*Cyamopsis tetragonoloba* (L.) Taub.], until recently remained a minor crop, seems destined to assume a larger role among the domesticated plants that cater the commercial needs of man. The discovery that guar seed endosperm could be a source of useful industrial gum brought this little known crop world recognition and paved its way to the crop of major prominence. Guar gum is highly mucilaginous and has several diversified uses. Guar gum and its derivatives are in exceeding demand in the world market. India earned a foreign exchange of Rs. 544 crores during 1997-98 and the export figures are expected to be increased considerably in the recent past. However, guar has so far received practically little attention by the researchers as far its genetic improvement is concerned. An attempt has, therefore, been made to systematically review the work carried out on its genetic upliftment with a particular reference to India.

Origin

In view of Gillete (1958), Africa was probably the centre of origin of *Cyamopsis*. The possibility that guar was domesticated very early in Africa and Arabia and then made its way to Indo-Pakistan sub-continent appears remote. The name of the cultigen in Arabic is *hindia*, which suggests it to be an Indian origin (Whistler and Hymowitz, 1979). Hymowitz (1972) proposed transdomestication concept to explain the origin of guar. According to him, a drought tolerant legume *Cyamopsis senegalensis* found in the semi-arid and Savanneh zone south of the Sahara from Senegal to Saudi Arabia, reached Indo-Pakistan sub-continent as flotsam in Arab-Indian trade. One of the major ancient items of trade between the Arabs and Indians were the horses. The Arabs must have boarded their ships in huge quantities of fodder to feed the horses. The plants of *C. senegalensis* probably were cut and taken along the ship as fodder. Since, the climatic conditions in the dry regions of the Indo-Pakistan sub-continent were favourable to *C. senegalensis*, seeds brought along by the Arabs, which germinated, could have easily formed the basis for further as *C. tetragonoloba*. The hypothesis on the origin of guar appears quite speculative. Therefore, biosystematic studies are needed to critically examine the relationship between the cultigen *Cyamopsis tetragonoloba* and the other species of *Cyamopsis* particularly, *C. senegalensis*.

Floral Biology

The guar flowers are borne in axillary racemes. The five unequal linear teeth type sepals form the calyx, while papilionaceous corolla has orbicular standard, oblong wing petal and long and broad keel petals. The flower colour in general is purplish to pink. However, lines with white colour flower are also available. There are 10 stamens in diadelphous condition in which apicular anther are present. Style is short and slender while stigma is head shaped (Chandersekharan and Ramakrishnan, 1928; Jafri, 1966; Menon *et al.*, 1968; Saini *et al.*, 1981).

Mode of pollination : Guar is strictly self-pollinated crop because of its cleistogamous nature. The extent of outcrossing has been found to vary from 0.3 to 7.92 per cent (Menon *et al.*, 1973a; Saini *et al.*, 1981; Chaudhary and Singh, 1986).

Crossing technique : Chowdhury *et al.* (1974) developed a crossing technique which is two to three times more effective than the usual direct emasculation and pollination procedure. By using this technique, the investigators could get a 7 per cent pod set as against 3 per cent pod set by the traditional crossing method.

Pollen

Staining of pollen grains in acetocarmine revealed 95 per cent pollen fertility in case of *C. serrata*, while it was about 90 per cent in *C. tetragonoloba*. Maximum pollen germination (64.3%) and pollen tube length (60.5μ) in case of *C. tetragonoloba* was obtained in 20 per cent sucrose (Tyagi, 1974). Interestingly, addition of 0.15 per cent boric acid to this medium enhanced the pollen germination to 84.7 per cent and tubes length to 208.4μ. The study also revealed that pollen grains of *C. tetragonoloba* required 25 per cent sucrose + 100 ppm boric acid + 250 ppm calcium nitrate and 0.5 per cent agar for best germination, whereas *C. serrata* pollen grains showed maximum germination at 35 per cent sucrose + 150 ppm boric acid + 300 ppm calcium nitrate + 1.0 per cent PEG + 0.5 per cent agar. This study further emphasized that sucrose (25%) produced the longest pollen tubes, while glucose (30%) gave the highest percentage of germination in the presence of calcium and boron. Moreover, proline, glutamic acid and chlorogenic acid promoted pollen germination and tube growth as well. However, aspartic acid increased germination but inhibited pollen tube growth.

CYTOGENETICS

The genus *Cyamopsis* D.C. is an old world genus (tribe Galegeae of family Papilionaceae) with four species viz., *C. tetragonoloba* L. Taub., *C. serrata* Schinz., *C. senegalensis* Guill and Perr and *C. dentata* Tarre. Saini *et al.* (1981) reported the morphological characters of three species barring *C. dentata*. The *C. tetragonoloba* (Syn. *C. psoralioides*) is a summer annual herb, has no wild forms at present anywhere except in Rawalpindi and Afghanistan which are considered as one of its habitats and it might have been introduced to India from these regions (Sarbhoy, 1977). The species *C. senegalensis* is found in semi-arid Savanneh zone of the Sahara from Senegal to Saudi Arabia.

The *C. serrata* and *C. dentata* have been reported from the semi-arid habitat of the Republic of South Africa, South-West Asia and Botswana. *C. serrata* and *C. senegalensis* are highly drought tolerant and resistant to diseases (Menon, 1973). Ayyangar and Krishnaswami (1933) observed seven pair of chromosomes in *C. tetragonoloba* (Syn. *C. psoraliodes*), confirmed by many investigators (Sen and Vidyabhushan, 1960; Vig, 1963a; Srivastava and Naithani; 1964; Arora *et al.*, 1985; Sandhu, 1988). Sandhu (1988) demonstrated that *C. serrata* also had seven pairs of chromosomes.

Srivastava and Naithani (1964) and Sarbhoy (1977)

studied the chromosome morphology in *C. tetragonoloba*, whereas Arora *et al.* (1985) first time compared the chromosome morphology of two wild species against cultivated species *C. tetragonoloba*. Sandhu (1988) had also restudied the chromosome morphology of *C. tetragonoloba* and *C. serrata*.

Karyotype analysis of three species of *Cyamopsis* showed an identity in chromosome number (2n=14) and similarity in morphology. All the chromosomes were found medianly or submedianly constricted. *C. serrata* has chromosome complement longer than that of *C. senegalensis* and *C. tetragonoloba*. The chromosome length in *C. tetragonoloba*, *C. serrata* and *C. senegalensis* ranged between 7.16 to 9.26μ ; 7.33 to 11.52 μ and 5.50 to 9.60 μ, respectively. The total form per cent (TF%) value of *C. tetragonoloba* (44%) showed the close proximity with that of *C. serrata* (45%); and *C. senegalensis* (45%) was also indicative of similarity in karyotypes among these species. All the three species have symmetrical karyotypes. The shift in centromere position seemed to have occurred in this genus with respect to three pairs of chromosomes (i.e. 2nd, 4th and 7th) only. However, it would be reasonable to assume that species *C. serrata* in view of slightly longer chromosomes may represent a comparatively primitive status in relation to rest of the species (Fig. 1).

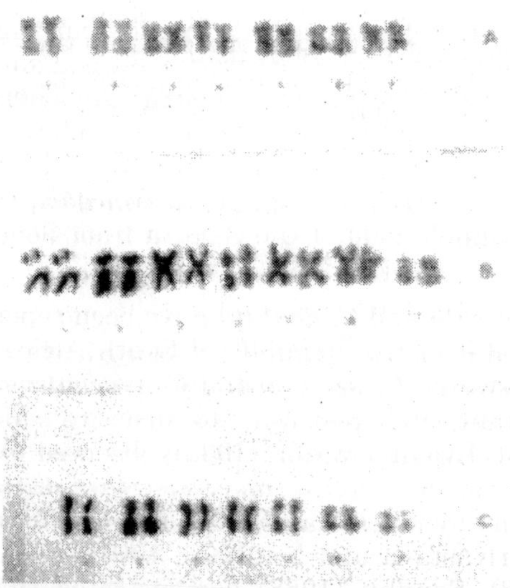

Fig. 1. Karyotypes of somatic chromosomes of three species of genus *Cyamopsis* (A) *C. tetragonoloba* (B) *C. serrata* and (C) *C. senegalensis*

Germplasm Collection and Evaluation

Enormous variability in guar germplasm is represented by branched or unbranched, hairy or smooth, straight or sickle shaped pod, pubescent or glabrus, determinate or indeterminate, regular or irregular pod bearing types. A wide range of genetic variability for plant height (78-226.6 cm), branching behaviour (0-22), pod length (9.0-12.5 cm), 100-seed weight (3.8-4.00 g), gum content (22.5-39.8%) was reported by Mittal *et al.* (1977) from a collection of more than 1500 accessions of guar. Moderate to high estimates of genotypic coefficient of variability for seed yield and most of its component characters of guar have also been reported by Sanghi *et al.* (1964), Mittal *et al.* (1969), Dass *et al.* (1973), Tikka *et al.* (1974), Tripathi and Lal (1975), Paroda *et al.* (1977), Dabas *et al.* (1982), Vijay and Leela (1989), Arora and Lodhi (1995) and Satyaparkash and Singh (2000). Publication of three volumes on the categorisation of germplasm collected from various parts of the country again proves vast variability for morphological traits of guar existing in India. The selections made from land races have also resulted in the release of improved cultivars of this crop (Singh and Sikka, 1955; Mittal *et al.*, 1963; Singh and Sharma, 1965; Saini *et al.*, 1982; Henry *et al.*, 1992; Bharodia *et al.*, 1993; Singh *et al.*, 1995).

Applied Research Findings

Efforts on selecting stable varieties of guar, performing well over wide range of environmental conditions have been taken up by a number of workers (Paroda and Mehrotra, 1976; Saini *et al.*, 1977; Saini *et al.*, 1979; Mehrotra *et al.*, 1980; Paroda and Rao, 1981; Rao *et al.*, 1982; Dangi, 1982; Henry and Daulay, 1983; Paroda *et al.*, 1983; Paroda *et al.*, 1990; Singh *et al.*, 1991; Lodhi *et al.*, 1996). Some promising and stable genotypes reported by the workers are : HG 365, HFG 314, RGC 197, FS 277, IC 9229/P3, HG 75, 2470 (12), HGS 43, GG1, Durgapura Safed, HG 182, RGC 165, No. 2, HFG 119, PLG 85, B 19-1-55, HFG 181 and HFG 244. Among these genotypes, the lines viz., HG 75, HG 365, FS 277, IC 9229/P3, GG-Durgapura Safed, HG 182, HFG 119, No. 2 and PLG 85 have been released for general cultivation and have still been performing well over a wide range of environments (Fig. 2). Some new plant types rich in harvest index, early and synchronous maturity and others with unbranched types have great applied and basic relevance (Figs. 2, 3 and 4).

Correlation studies conducted by various workers in guar (Sanghi and Sharma, 1964; Mittal and Thomas, 1969; Tikka, 1975; Menon 1973a; Nath, 1979; Nath and Saini, 1980; Bhardwaj *et al.*, 1981; Sindhu *et al.*, 1982; Singh *et al.*, 1988; Vijay, 1988) indicated that seed yield was positively correlated with number of pods per plant, pod length, number of clusters per plant, number of branches per plant and 100-seed weight. Based on these studies, it may be concluded that a model plant for higher seed yield must possess few branches, high number of clusters and pods with bolder seeds and long peduncle.

Positive correlations were observed between grain yield and gum percentage, whereas negative correlations were obtained between seed weight and gum percentage (Menon *et al.*, 1970; Mittal *et al.*, 1971). Positive correlation was also reported between endosperm percentage and percentage of gum in endosperm (Menon *et al.*, 1970; Lal and Gupta, 1977).

The selection of parents for hybridization based on individual attribute may not be as fruitful as based on certain group of characters particularly under situations when the character for improvement is complex and is quantitative in nature.

The usefulness of a hybrid programme depends on the extent of heterosis available for economic traits, it is, therefore, desired to have an estimation on the magnitudes of heterosis for important traits. Saini and Jhorar (1979) reported that out of 21 crosses, 17 had significant heterosis for seed yield over the better parent (140.27%). Saini and Jhorar (1979) and Hooda (1981) reported range from –26.22 to 36.61% for heterosis over better parent for seed yield, while Chaudhary *et al.* (1981) reported the same from 9.43 to 95.90. These reports, therefore, clearly reveal that a wide range of heterosis exists in this crop, nevertheless heterosis breeding has its own limitations in guar as till date no stable source of male sterility is available in this arid legume.

The combining ability analysis conducted by Jhorar *et al.* (1985), Jhorar *et al.* (1988), Hooda *et al.* (1990), Jhorar *et al.* (1990), Arora *et al.* (1996) and Arora and Lodhi, 1998 indicated that both additive and non-additive genetic variances were important in the expression of all the characters. However, the former appeared to be of greater importance for seed yield and 100-seed weight.

Most of the pods at the time of harvesting remain green in guar, leading to considerable loss due to its indeterminate growth habit. Such plants when harvested result in

Fig. 2. HG 365 - An early maturing variety of guar

Fig. 3. New plant type of guar

Fig. 4. Improved unbranched type of guar

considerable number of blackened seeds which further reduce the seed quality. The determinate mutant in guar thus holds promise for early maturity and high seed yield as well. Saini *et al.* (1981) isolated natural mutant for determinate characters in FS 277 and HFG 257 and categorised determinate habit in guar into three groups viz., (i) termination of growth by leaf emergence, (ii) termination with inflorescence or (iii) the main stem finally terminated into inflorescence. As such, these were termed as leaf, inflorescence and stem determinate, respectively (Figs. 5, 6 and 7). They also observed that time of termination had direct impact on the plant height, earliness and peduncle length. These indeterminate mutants may prove useful as such and in hybridization programme as well, aiming breeding for earliness and photo-insensitivity, enabling the possibility of two crops in a year under assured irrigation facilities.

Basic Research Findings

The information on various genetic aspects of important plant traits would be useful in shaping out breeding programme for yield enhancement. The inheritance studies indicated that pollen fertility was monogenically dominant over sterility. The foliage colour was controlled by single gene and was dominant for dark green foliage over the light green foliage. The F_1s between the long and short podded type were intermediate and pod length was controlled by several gene pairs (Dabas, 1975). The inheritance of seven qualitative characters studied by Sahoo and Gill (1976), led to the gene symbols viz., H/h (hairy/non-hairy), Y/y (green/yellow leaflet), N/n (narrow/normal leaflet), B/b (normal/bushy plant type) and R/r (flat/round seed). Dabas (1975) reported unbranched monogenic dominance over branched one, while Chaudhary and Lodhi (1981) and Singh *et al.* (1990) observed complementary gene effects for branching behaviour. On the other hand, Saharan *et al.* (1999) found monogenic inheritance behaviour of branching. Serrated leaf margin was reported dominant over the smooth leaf margin and ratio of 13:3 was observed, whereas 3:1 ratio was observed for pubescent versus glabrous leaf surface (Singh *et al.*, 1989). On the other hand, Saharan *et al.* (1999) was of the view that inheritance for leaf serration was a monogenic trait. The mode of inheritance of nine qualitative characters was reported dominant over others for instance, hairy over non-hairy, branched over unbranched, plucked leaf surface over non-plucked, broad leaf over narrow leaf, smooth leaf margin over serrated, purple flower colour over white and curved pod

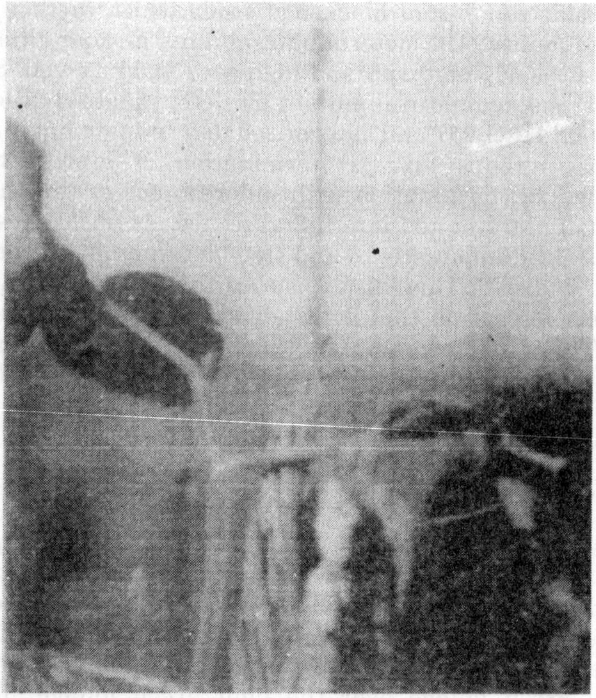

Fig. 5. Leaf determinate mutant.

Fig. 6. Inflorescence determinate mutant.

Fig. 7. Stem determinate mutant.

over non-curved (Saharan *et al.,* 1999). Complementary gene effects for pod bearing habit and duplicate for non-fleshy pod over fleshy pod were also reported by these workers.

High estimates of heritability and genetic advance have been observed for many traits by a number of workers (Sanghi *et al.*, 1964; Dass *et al.*, 1973; Tripathi and Lal, 1975; Chaudhary and Singh, 1976; Reddy and Gupta, 1984; Singh *et al.*, 1989; Lokesha and Shiv Shankar, 1990; Chaudhary *et al.*, 1991). The genetics of seed yield and its components indicated that these traits were in general governed by non-additive genetic variances. However, additive genetic component appeared to play significant role in the inheritance of characters like, days to first bloom, plant height, branches per plant, pods per plant and seeds per pod (Nath, 1979; Jhorar and Solanki, 1983; Jhorar *et al.*, 1984; Jhorar *et al.*, 1988; Singh *et al.*, 1990; Hooda *et al.*, 1991; Arora *et al.*, 1993, 1999).

Breeding for Quality Traits

Gum and protein contents in guar seed are of prime importance from the industrial view point. Considerable variability for these important traits exists in the available germplasm resources (Menon *et al.*, 1970; Dass *et al.*, 1973). However, significant strides in enhancement of gum and protein contents in the seed have not been made due to certain inconvenient situations. For instance, positive correlation was observed between seed yield and percentage gum content (Menon *et al.*, 1970; Mittal *et al.*, 1971; Lal and Gupta, 1977) but there was negative correlation between seed weight and gum percentage, which may hamper breeding for bold seededness. Likewise seed colour did not show any bearing on the seed gum content. Genetic analysis revealed that both additive and non-additive gene effects were operating in the expression of this industrial component (Dabas *et al.*, 1980; Jhorar *et al.*, 1985, 1989; Singh *et al.*, 1990). Thus, more concrete efforts are required to be taken up to improve these two important traits in guar.

Disease Resistance

Although an array of diseases infect guar crop, however, bacterial blight (caused by *Xanthomonas oxonopodis)*, Alternaria leaf spot and powdery mildew are the serious diseases, which appear almost every year and seriously threatened guar production in many pockets. The most practical and feasible approach to this problem is development of varieties having inbuilt resistance to these diseases so as to check serious outbreak of disease and considerable yield losses. The existing literature provides a wealth of information on source of diseases resistance based on screening of germplasm and improved lines (Mittal *et al.*, 1977; Gandhi *et al.*, 1978; Gupta *et al.*, 1998). However, the inheritance of bacterial blight and Alternaria leaf spot is not fully understood. Singh *et al.* (1995) and Saharan *et al.* (1999) observed that resistance for blacterical blight was dominant trait, whereas in case of Alternaria leaf spot reverse was the case. Many studies have also pointed out that additive, dominance and epistatic gene interactions played significant role towards contributing resistance in guar. From these reports, however, it is not clear whether two or more of the genes reported conferring resistance to these diseases are identical. The genetics of disease resistance would be further clarified if the individual genes responsible for resistance are placed onto an accurate genome map for guar.

Different biochemical parameters imparting resistance have also been studied by Singh *et al.* (1981), Sindhu and Prashar (1981), Singh *et al.* (1989), Singh *et al.* (1990) in this crop. Phenol and peroxidase enzyme played significant role in imparting the resistance towards main diseases in this crop. The biochemical traits showed significant and negative correlation with wilt, bacterial blight and Alternaria leaf spot (Singh *et al.*, 1990). The genetics of phenol and peroxidase enzyme indicated the role of additive gene in their inheritance (Singh *et al.*, 1991, 1994).

MUTATION BREEDING

Mutation breeding is a powerful tool for enhancing variation in a crop like guar, where exploitable and useful genetic variability is meagre. Moreover, creation of variability in this crop through hybridization is very difficult and cumbersome owing to small, delicate flower structures resulting in low percentage of seed setting in the manually hybridized buds. Efforts have, therefore, been made to induce a purposeful variability in guar by various workers through various mutagens.

(i) Physical Mutagens

Vig (1965, 1969) made first successful attempt to induce mutations in guar through physical mutagens, and reported the occurrence of reciprocal translocation in the 20 kR gamma irradiated population of guar which exhibited low fertility. Trisomic plants were observed by Singh (1972) in gamma irradiated M_2 progenies. Lather and Chowdhury (1972) studied the effects of irradiation on morphology and cytology of two varieties of guar using 10, 20, 40, 60, 80, 100, 150 and 200 kR doses of gamma rays and found that germination percentage and seedling survival decreased with increased doses. The dose of 200 kR proved to be lethal leading to complete failure in seed germinations in both varieties. Pollen fertility also decreased with the increasing doses resulting in considerable yield loss. The varietal differences for sensitivity to radiations for various characters were also observed. On the contrary, Chaudhary *et al.* (1973) observed stimulatory effects on yield, protein and gum contents in the M_2 populations following seed irradiation with 2, 5, 10, 15 and 20 kR doses of gamma rays. The possible cause for enhancement in the expression of characters could be the low doses of irradiation proving stimulatory effect. Chowdhury *et al.* (1975) studied the effects on induced variability in M_2 following irradiation doses ranging from 10-25 kR gamma rays on two varieties of guar for six quantitative characters and found considerable variations in the irradiated populations. Early flowering and determinate mutants were developed by Singh *et al.* (1981) by gamma irradiation. Rao and Rao (1982) studied the effect of 0-60 kR doses of X-rays on seeds of cv. 'Pusa Navbahar' and found that low doses were stimulatory, while high doses proved inhibitory, however, the inhibitory effects were decreased with the advancement of generation. They also isolated a stabilized early flowering mutant with more number of pods at 10 kR dose Co^{60} gamma ray.

(ii) Chemical Mutagens

Chlorophyll deficient mutants and other mutants having profuse vegetative growth were observed by Gohal *et al.* (1970) in M_1 following seeds treated with EMS. Swamy and Hashim (1980) isolated unbranched and regular pod bearing mutants by treating the seeds with hydroxyl amine. They also isolated mutants with changed leaf structure, growth habit and pod size in the population of seeds treated with EMS or hydrazine hydrate and also observed that some of the mutants had pleiotropic effect i.e., extensive branching and late flowering. Some mutants with changes in seed colour from normal violet to light grey or light brown were also induced. Rao *et al.* (1982)

observed determinate and spreading variants in cv. 'Pusa Navbahar' which soaked in 200, 400 and 600 ppm Ketazin and 1000, 2000 and 3000 ppm saturn for 12 and 24 h.

(iii) Physical and Chemical Mutagens

Induction of high yielding mutants in cv. 'PLG 143' and 'Suvidha' seeds exposed to 80 and 100 kR gamma rays and treated with aqueous solutions of 0.1-0.3% EMS or 0.01-0.03% solution of N-methyl-N-nitrosourea (NMU) either alone or in various combinations have been observed. The resultant M_2 mutants had long pods, increased number of pods and early maturity. These mutants also had increased yields and gum contents in the endosperm or whole seed. Basha and Rao (1988) induced heterophylly in guar by exposing the seeds to gamma rays and sodium azide treatment and reported that combined treatment of 35 and 45 kR with 100 and 200 ppm sodium azide could produce two and four leaflet instead of three.

(iv) Effect of Hybridization and Irradiation on Induction of Variability

The comparison of nature and magnitude of variability among the unirradiated F_1s and F_2s and the irradiated F_1M_1s and F_2M_2s revealed that irradiation caused earliness in flowering and maturity as well; and increased 100-seed weight. The phenotypic and genotypic coefficients of variability increased for pod length, seeds per pod, protein and gum content and reduced for days to flowering, maturity, plant height, clusters and seed yield per plant in both F_1M_1 and F_2M_2 irradiated populations. The mean values reduced, whereas coefficient of variation increased in majority of the crosses in F_2M_2 population (Arora and Lodhi, 1995).

Combining ability analysis for protein and gum content in unirradiated and irradiated diallel populations of guar revealed that the parent HG 75 in unirradiated F_1 and F_2 populations became poor combiner in F_1M_1 and F_2M_2 irradiated populations. Contrary to it, parent PLG 85 which was poor combiner in unirradiated F_1 proved good combiner in F_1M_1. This showed that the effect of irradiation varied with the genotypes due to difference in their genetic background. Many F_1 and F_2 crosses which were irradiated showed high sca effects for protein and gum content. A change in magnitude and direction of the correlation coefficients with seed yield was observed for plant height, days to flowering and maturity, branches per plant, 100-seed mass and seeds per pod from F_1M_1 to F_2M_2.

Future Challenges in Mutation Breeding

It is obvious from the foregoing discussion that the lower doses of mutagens induced useful variability compared to those of higher doses. Certain mutants carrying one or two useful attributes have also been isolated. Studies on creation of genetic variability following hybridization and irradiation have opened up new vistas and thus, more planned studies on these aspects can be initiated to obtain more desirable mutants in guar.

Polyploidy

Clusterbean is a diploid with 2n=14. Two wild species, namely *C. serrata* and *C. tetragonoloba* are also diploid with chromosome no. 2n=14. However, autotetraploids have been synthesized and studied in detail in the cultivated spp. *C. tetragonoloba*. The autotetraploids have been characterized by comparatively slow growth during early stages of development, thickened leaves, fewer stomata and larger pollen grains. The autotetraploids show severe irregularities in meiosis resulting in reduced pollen and seed fertility. Quadrivalents, trivalents, univalents and laggards have been observed in induced tetraploids (Sen and Vidyabhushan, 1960; Vig, 1963b; Biswas and Bhattacharya, 1971; Lobana and Verma, 1972; Sahoo and Gill, 1976). A mixoploid plant resulted from aqueous colchicine treatment showed 2n=26 and 2n=28 chromosomes in the PMC's of the same plant. Sen and Vidyabhushan (1960) also detected aneuploids having 2n=29 chromosomes which were less fertile and also differed in several aspects from tetraploids. Use of induced polyploids for improvement in guar appears to be of rather limited value, however, their use in developing interspecific hybrids can be of some use.

Interspecific Hybridization

Sandhu (1988) while studying the details of hybridization between two species of guar (*C. tetragonoloba* and *C. serrata*) observed that the hybridization though a conventional method could be a complete failure. The pollen grains failed to stick to the incompatible stigma probably due to pollen rejection by the stigma. The various other approaches like, bud pollination, amputation of stigma and style, use of organic solvents and mentor recognition of pollen failed to overcome the stigmatic incompatibility barrier due to the failure of pollen germination on the foreign stigma. The crossing programme modified with the help of pollen germination nutrient, resulted in pollen germination on the incompatible stigma. The number of crossed pistils showed an increase in length and width but without any seed development. Similarly, the application of nutrient media on the amputated stylar surface and during bud pollination led to the pollen germination but the pollen tube growth was not normal. *In vitro* pollination technique overcame the problem of pistil abscission to a great extent. However, the pistil growth was still without the seed formation. During pollen tube growth studies in the crossed pistils, the number of pollen tubes, heavy deposition of callose along the pollen tubes and the bulbous pollen tube, etc. were noticed in the ovarian tissues between *C. tetragonoloba* x *C. serrata*. Ovule histology revealed that unfertilized ovules were present in the crossed pistils. The observed increase in pistil length and width may be due to stimulatory effects of pollen tubes in the ovarian tissues. Earlier attempts to make crosses among *C. tetragonoloba*, *C. serrata* and *C. senegalensis* also failed, mainly due to the reasons specified by Sandhu (1988). Tissue culture techniques like somatic protoplast hybridization between these spp. may be tried in future to develop interspecific hybrids in the genus *Cyamopsis* or recombinant DNA technology may also be useful to transfer the important genes from wild species to the cultivated one.

Biotechnological Tools

The use of molecular marker to enhance and accelerate selection technique has been reviewed elsewhere. The potential of recombinant DNA technology for the development of improved crop varieties has also been explored and it reveals that conventional plant breeding has still much to contribute. However, for accelerated and targeted transfer of genes for characters such as disease resistance, resistance to herbicides though distant crosses, genetic engineering may be important in association with plant breeding. At present, the major problem is the lack of cloned DNA representing genes for characters of economic importance for use of guar breeders. The use of RFLPs for gene mapping though expensive, could provide a useful selection tool for the breeders. Agrobacterium-mediated gene transfer and direct gene transfer into protoplasts are methods which have yielded transgenic plants. Such developing technologies are certain to find significant application in guar breeding. The initial studies conducted on these aspects at CCS

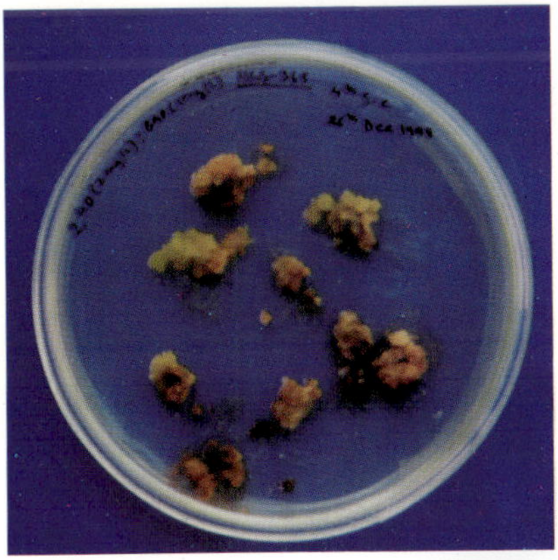

Callus maintenance of genotype HG -365

Embryo explant of genotype RGC-936

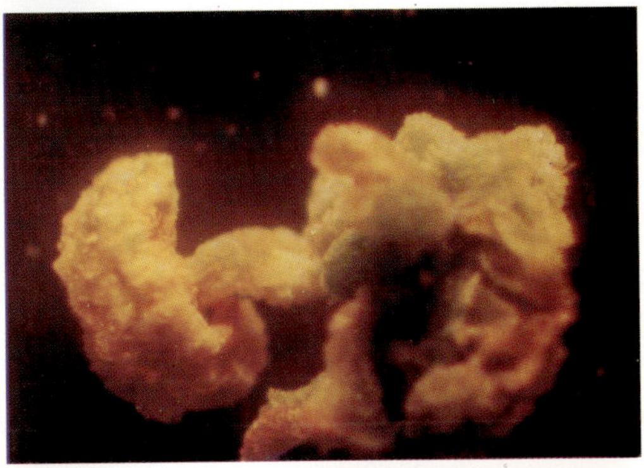

Direct shoot morphogenesis

Root initiation

Plate - 1. Tissue culture studies in guar

HAU, Hisar have indicated that in guar embryo and shoot tip explant exhibited maximum callus induction (Plate 1). Kanamycin concentration of 200 mg/l has been standardized for growth of transgenic calli cells. Agrobacterium infection in terms of GUS has been observed genotypic and tissue independent (Deepak Prem, 1999).

FUTURE THRUST

As in several self-pollinated species, in guar also varietal improvement methods based on pedigree, bulk pedigree, backcross and multiple-crossing techniques have been useful in recombining simply inherited characters such as disease resistance, seed size and colour, and maturity duration. These improvement methods have not, however, been very efficient in improving the quantitatively inherited traits like, seed yield. Selection among lines derived from advanced bulk populations appears more efficient than the pedigree method in breeding for seed yield. Local landraces could be utilized to derive adapted lines for use in breeding programmes. Single plant selection is generally ineffective for yield, hence, selection must be done on a progeny basis. For effective use, there is great need in the future for cleistogamous flower morphology to ensure near-complete self-pollination to maintain the purity of lines.

The major limitations in multilocational testing of advanced material are, the high coefficient of variability, high genotypic x environment interactions, and the adaptability to different production systems. Selection of suitable sites and production systems, and characterization of test environments are vital for precise testing. Future progress will depend on streamlining procedures to ensure precise testing and efficiency in selection. However, more work is needed to quantify on existing production systems and identifying niches of the cropping systems and environments where new short-duration plant types can be used. It is in these environments and cropping systems that the testing of appropriate genotypes need to be done.

For commercial production in multiple cropping systems, the genotypes maturing in less than 90 days are required. These extra early and early-maturing genotypes may prove more amenable to fitting in certain environmental windows. Such genotypes would also be relatively less sensitive to photoperiod and temperature interactions. These short-statured determinate cultivars are suitable for mechanized commercial cultivation. Improvement in harvest index and growth rate will contribute to increased grain yield within a maturity duration in any crop. In short-duration guar introgression from wild *Cyamopsis* species for rapid growth rate, salinity tolerance, etc. should make future research considerations.

A new ideotype visualized in guar, would be with short-duration maturity, photoperiod-insensitive, and determinate growth habit, with an elevated harvest index (>30%) and faster growth rate. Modifications of this ideotype could be used for different purposes and in different environments.

SUMMARY

Clusterbean (*Cyamopsis tetragonoloba* (L) Taub.) commonly known as guar is a drought hardy crop of the Indian Arid Zone. The discovery that guar seed endosperm could be source of useful industrial gum brought this little known crop world recognition and started it on its way to major crop prominence. This review highlights progress made in stabilizing guar production through breeding. Several high yielding varieties have been released by different states. Considerable progress has also been made in identifying germplasm.

Progress made during last 50 years through breeding is discussed. Interspecific hybridization and newly developed biotechnological tools could play an important part in improving the effectiveness of conventional breeding programmes.

REFERENCES

Arora, R. N. and Lodhi, G P., 1995. Genetic variability in guar following hybridization and irradiation. *Indian J. Pl. Genet. Res.* **8** (2) : 201-208. Special Issue (3).

Arora, R. N., Lodhi, G. P. and Singh, J. V., 1996. A comparison of four diallel analyses in clusterbean [*Cyamopsis tetragonoloba* (L.) Taub.]. *Forage Res.* **22** (1) : 45-50.

Arora, R. N., Lodhi, G. P. and Arora, S. K., 1993. Combining ability analysis for quality characters in normal and irradiated diallel population of guar [*Cyamopsis tetragonoloba* (L.) Taub.] *Forage Res.* **19** (3&4) : 299-305.

Arora, R.N. and Lodhi, G.P., 1998. Combining in ability studies for yield and its components in clusterbean. *Indian J. Genet.* **58** (4) : 525-528.

Arora, R. N., Lodhi, G. P., Singh, J. V., Kishore, C. and Jhorar, B. S., 1999. Genetic analysis for grain yield and its component characters in F_2 population of guar. *Ann. Biol.* **15**(1) : 45-49.

Arora, R. N., Sareen, P. K., Saini, M. L., Paroda, R. S. and Singh, J. V., 1985. Karyotype analysis in three species of genus *Cyamopsis*. *Indian J. Genet.* **45** (2) : 302-309.

Ayyangar, G. N. P. and Krishnaswami, N., 1933. A note on the chromosome number in clusterbean (*Cyamopsis psoralioides* DC). *Indian J. Agri. Sci.* **3** : 934-935.

Basha, S. K. and Rao, P. G., 1988. Gamma rays and sodium azide induced heterophylly of bhindi and clusterbean. *J. Nucl. Agri. Biol.* **17** : 133-136.

Bhardwaj, R. P., Mathur, M. L. and Mathur, R., 1981. Inter-relationship among yield and yield components in grain varieties of *Cyamopsis tetragonoloba* (L.) Taub. *Madras Agri. J.* **68** (9) : 594-598.

Bharodia, P. S., Zaveri, P. P., Kher, H. R., Pasel, M. P. and Chaudhari, D. N., 1993. GAUG 34 – a high yielding variety of clusterbean. *Indian Fmg.* **43** (9) : 31-33.

Biswas, A. K. and Bhattacharya, N. K., 1971. Induced polyploidy in legumes. *Cyamopsis psoralioides* D.C. *Cytol.* **36** : 469-479.

Chandersekharan, S. N. and Ramakrishnan, T. S., 1928. Botany of some useful plants. *Madras Agri. J.* **16** : 5-11.

Chaudhary, B. D. and Singh, V. P., 1976. Studies on variability, correlation and path analysis in guar. *Forage Res.* **2** : 97-103.

Chaudhary, B. S. and Lodhi, G. P., 1980. Correlation and path analysis for fodder traits in guar. *Forage Res.* **6** (2) : 129-134.

Chaudhary, B. S. and Lodhi, G. P., 1981. The inheritance of five quantitative characteristics in clusterbean (*Cyamopsis tetragonoloba*). *Euph.* **30** (1) : 161-166.

Chaudhary, B. S. and Singh, V. P., 1986. Extent of outcrossing in guar [*Cyamopsis tetragonoloba* (L.) Taub.]. *Genet. Agrar.* **34** : 59-62.

Chaudhary, B. S., Lodhi, G. P. and Arora, N. D., 1981. Heterosis for grain yield and quality characters in clusterbean. *Indian J. Agri. Sci.* **51** (9) : 638-642.

Chaudhary, B. S., Battan, K. R. and Lodhi, G. P., 1991. Studies on variability, heritability and genetic advance in clusterbean. *Indian J. Agri. Res.* **25** (1) : 14-20.

Chaudhary, M. S., Ram, H., Hooda, R. S. and Dhindsa, K. S., 1973. Effect of gamma irradiation on yield and quality of guar [*Cyamopsis tetragonoloba* (L.) Taub.]. *Ann. Arid Zone* **12** : 19-22.

Chaudhary, B. S., Paroda, R. S. and Solanki, K. R., 1974. A new crossing technique in clusterbean [*Cyamopsis tetragonoloba* (L.) Taub.]. *Curr. Sci.* **43** : 456-59.

Chowdhury, R. K., Chowdhury, J. B. and Singh, R. K., 1975.

Induced polygenic variability in clusterbean. *Crop Imp.* **2** : 17-24.

Dabas, B. S., 1975. Studies on inheritance of quantitative characters and gum content in guar [*Cyamopsis tetragonoloba* (L.) Taub.]. *Ph.D. Thesis* (Unpublished), IARI, New Delhi.

Dabas, B. S., Mittal, S. P. and Arunachalam, V., 1982. An evaluation of germplasm accessions in guar. *Indian J. Genet.* **42** (1) : 56-59.

Dabas, B. S., Mittal, S. P., Singh, H. B. and Swarup, V., 1980. Inheritance of endosperm percentage of gum content in guar. *Indian J. Genet.* **40** (1) : 8-12.

Dangi, K. S., 1982. Genotype-environment interactions for seed yield in clusterbean (*Cyamopsis tetragonoloba*). *Thesis Abstracts* **8** (3) : 251, Haryana Agric. Univ., Hisar.

Dass, S., Arora, N. D. and Singh, V. P., 1973. Heritability estimates and genetic advance for gum and protein content along with seed yield and its components in clusterbean [*Cyamopsis tetragonoloba* (L.) Taub.]. *Haryana Agric. Univ. J. Res.* **3** : 14-19.

Deepak, Prem, 1999. Callus induction regeneration and agrobacterium mediated transformation of clusterbean (*Cyamopsis tetragonoloba* L.). *M. Sc. Thesis*, CCSHAU, Hisar.

Gandhi, S. K., Saini, M. L. and Jhorar, B. S., 1978. Screening of clusterbean genotypes for resistance to leaf spot caused by *Alternaria cyamopsis*. *Forage Res.* **4** (2) : 169-172.

Gillete, J. B., 1958. Indigofera (Microcharis) in tropical Africa with the related genera Cyamopsis and Rhyncho tropis. *Kew Bulletin*, Additional Series. **1** : 1-166.

Gopal, M. S., Kalia, H. R., Dhillon, M. S. and Negi, K. S., 1970. Effect of ethyl methane sulphonate on the mutation spectrum in guar. *Indian J. Hered.* **3** : 51-54.

Gupta, P. P., Arora, R. N. and Gandhi, S. K., 1998. Diseases of *kharif* fodder crops and their management. pp. 373-404. In : *Diseases of field crops and their management* (Ed. T. S. Thind). National Agri. Technol. Inform. Centre, Ludhiana.

Henry, A. and Daulay, H. S., 1983. Genotype x environment interactions for seed yield in clusterbean. *Indian J. Agri. Sci.* **53** (6) : 467-468.

Henry, A., Daulay, H. S. and Bhati, T. K., 1992. 'Maru Guar' promising clusterbean for arid region. *Indian Fmg.* **42** (6) : 24-25.

Hooda, J. S., Saini, M. L. and Singh, J. V., 1990. Combining ability studies in clusterbean. *Haryana Agric. Univ. J. Res.* **20** (1) : 28-34.

Hooda, J. S., Saini, M. L. and Singh, J. V., 1991. Genetics of quantitative characters in clusterbean. *Indian J. Trop. Agri.* **9** (2) : 101-109.

Jafri, S. M. H., 1966. The flora of Karachi, Karachi Book Corporation.

Jhorar, B. S. and Solanki, K. R., 1983. Genetic analysis for plant height and number of branches for clusterbean. *Forage Res.* **9** (2) : 147-150.

Jhorar, B. S., Solanki, K. R. and Jatasra, D. S., 1984. *Crop Imp.* **11** (2) :103-105.

Jhorar, B. S., Solanki, K. R. and Jatasra, D. S., 1985. Combining ability analysis of seed yield in clusterbean under different environments. *Cuban J. Agri. Sci.* **19** (1) : 113-119.

Jhorar, B. S., Solanki, K. R. and Jatasra, D. S., 1988. Combining ability analysis of kernel weight in clusterbean under different environments. *Indian J. Agri. Res.* **22** (4) : 188-192.

Jhorar, B. S., Solanki, K. R. and Dhukia, R. S., 1989. Inheritance of protein content in clusterbean under different environments. *Ann. Agri. Res.* **10** (2) : 115-119.

Jhorar, B. S., Solanki, K. R. and Sharma, S. C., 1990. Genetic architecture of protein content in clusterbean under different environments. *Canad. Res. Ann.* **6** : 19-21.

Lal, B. M. and Gupta, O. P., 1977. Studies on galactomannans in guar and some correlations for selecting genotypes rich in gum content. *Proc. First Guar Res. Workshop*, CAZRI, Jodhpur, pp. 124-130.

Lather, B. P. S. and Chowdhury, J. B., 1972. Studies on irradiated guar. *Nucl.* **15** : 16-22.

Lobana, K. S. and Verma, K. K., 1972. Induced tetraploidy in clusterbean. *Indian J. Genet.* **32** : 153-155.

Lodhi, G. P., Singh, J. V., Singh, V. P., Ram, H. and Kumar, Naresh, 1996. Genotype environment interaction in clusterbean. *Forage Res.* **21** (4) : 217-219.

Lokesha, R. and Shiv Shankar, G., 1990. Analysis of genetic variability and character association in clusterbean. *Mysore J. Agri. Sci.* **24** (3) : 316-320.

Mehrotra, N., Malik, D. S. and Chaudhary, B. D., 1980. Studies in stability in clusterbean [*Cyamopsis tetragonoloba* (L.) Taub.] for seed yield in rainfed conditions. *Haryana Agric. Univ. J. Res.* **10** (1) : 77-80.

Menon, U., 1973. A comprehensive review of crop improvement and utilization of clusterbean [*Cyamopsis tetragonoloba* (L.) Taub.]. *Monograph Series 2,* Dept. of Agri., Rajasthan, pp. 51.

Menon, U., Dube, M. M. and Bhargava, P. D., 1970. Gum content variations in guar [*Cyamopsis tetragonoloba* (L.) Taub.]. *Indian J. Hered.* **2** : 55-58.

Menon, U., Dube, M. M. and Chandola, R. P., 1973a. Genetic variability and correlation studies in vegetable clusterbean [*Cyamopsis tetragonoloba* (L.) Taub.]. *Rajasthan J. Agric. Sci.* **4** : 67-73.

Menon, U., Dube, M. M., Kala, K. C. and Chandola, R. P., 1973b. Path coefficient analysis in guar [*Cyamopsis tetragonoloba* (L.) Taub.]. *Indian J. Hered.* **5** : 7-10.

Menon, U., Radhare, N. S. and Bhargava, P. D., 1968. Pollen studies in guar [*Cyamopsis tetragonoloba* (L.) Taub.]. *Indian J. Palynol.* **4** : 51-53.

Mittal, S. P. and Thomas, T. A., 1969. Correlations and selection indices in improvement of seed yield in guar. *Indian J. Genet.* **29** : 1077.

Mittal, S. P., Dabas, B. S., Thomas, T. A. and Chopra, D. P., 1977. Assessment of genetic stocks of guar [*Cyamopsis tetragonoloba* (L.) Taub.]. *Proc. First ICAR Guar Res. Workshop*, CAZRI, Jodhpur, pp. 58-68.

Mittal, S. P., Thomas, T. A. and Srivastava, G., 1963. "Pusa Navbahar" guar goes to the market earlier. *Indian Fmg.* **13** (8) : 13.

Mittal, S. P., Thomas, T. A., Dabas, B. S. and Lal, S. M., 1971. Gum content as related to seed yield and other characters in guar. *Indian J. Genet.* **31** : 228-232.

Mittal, S. P., Swarup, V., Kohli, M. M. and Singh, H. B., 1969. Variability in guar. *Indian J. Genet.* **29** : 98-103.

Nath, R., 1979. Inheritance studies in clusterbean [*Cyamopsis tetragonoloba* (L.) Taub.] with particular reference to seed yield and its quality. *Thesis Abst.* **5** (4): 277, Haryana Agril. Univ., Hisar.

Nath, R. and Saini, M. L., 1980. Correlation and path coefficient analysis in clusterbean [*Cyamopsis tetragonoloba* (L.) Taub.]. *Forage Res.* **6** (1) : 1-6.

Paroda, R. S. and Rao, G. V. S., 1981. Genotype x environment interactions for seed yield in clusterbean. *Forage Res.* **7** (2) : 169-172.

Paroda, R. S. and Mehrotra, N., 1976. An investigation of genotype/environment interactions for green fodder yield in clusterbean [*Cyamopsis tetragonoloba* (L.) Taub.]. *Genet. Agrar.* **30** : 191-200.

Paroda, R. S., Singh, J. V. and Saini, M. L., 1983. Phenotypic stability for seed yield in clusterbean (*Cyamopsis tetragonoloba*). *Crop Imp.* **10** (1) : 49-52.

Paroda, R. S., Singh, J. V., Saini, M. L., Arora, R. N. and Sharma, S. C., 1990. Stability studies for seed yield in unbranched genotypes of clusterbean. *Forage Res.* **16** (2) : 150-152.

Paroda, R. S., Saini, M. L. and Jhorar, B. S., 1977. Genetic improvement in guar: problems and perspective. *Proc. First ICAR Guar Res. Workshop*, CAZRI, Jodhpur, pp. 27-42.

Rao, G. V. S. and Paroda, R. S., 1982. Factor analysis in clusterbean [(*Cyamopsis tetragonoloba* (L.) Taub.]. *Theo. Appl. Genet.* **62** (3) : 273-276.

Rao, G. V. S., Paroda, R. S. and Gupta, A., 1982. Genotype x environment interactions for protein content in clusterbean. *Forage Res.* **8** (1) : 55-58.

Rao, S. and Rao, D., 1982. Studies on the effect of X-irradiation on *Cyamopsis tetragonoloba* (L.) Taub. *Proc. Indian Nat. Sci. Acad. (Biol. Sci.)* **48** : 410-415.

Rao, S. R. M., Murty, P. K. and Rao, D., 1982. Note on determinate and spreading variants in clusterbean. *Curr. Sci.* **51** (19) : 945-956.

Reddy, J. N. and Gupta, S. C., 1984. Variability parameters in guar. *Madras Agri. J.* **71** (6) : 361-364.

Saharan, M. S., Saharan, G. S. and Singh, J. V., 1999. Inheritance of leaf margin and branching behaviour in clusterbean [*Cyamopsis tetragonoloba* (L.) Taub.]. Paper presented in *Nat. Sympo. on "Role of resistance in extensive agriculture"* held at IARI, New Delhi.

Saini, M. L. and Jhorar, B. S., 1979. Heterosis in clusterbean. *Forage Res.* **5** (1) : 5-11.

Saini, M. L., Jhorar, B. S. and Solanki, K. R., 1977. Genotypic/ environment interactions for seed yield in clusterbean. *Indian J. Agri. Sci.* **47** : 345-347.

Saini, M. L., Jhorar, B. S. and Solanki, K. R., 1979. Genotype x environment interaction for green fodder yield in clusterbean. *Forage Res.* **5** (2) : 185-188.

Saini, M. L., Arora, R. N. and Paroda, R. S., 1981. Morphology of three species of genus *Cyamopsis*. *Guar Newsl.* **2** : 7-11.

Saini, M. L., Paroda, R. S. and Arora, S. K., 1982. Genotype x environment interaction for gum content in clusterbean (*Cyamopsis tetragonoloba*). *Genet. Agrar.* **36** (1/2) : 57-62.

Sandhu, H. S., 1988. Interspecific hybridization studies in genus *Cyamopsis*. Unpublished Ph. D. Thesis, CCSHAU, Hisar.

Sanghi, A. K. and Sharma, S. K., 1964. Correlations in guar. *Indian J. Genet.* **24** : 282-285.

Sanghi, A. K., Bhatnagar, M. P. and Sharma, S. K., 1964. Genotypic and phenotypic variability in yield and other quantitative characters in guar. *Indian J Genet.* **24**: 164-167.

Sarbhoy, R. K., 1977. Cytogenetical studies in *Cyamopsis tetragonoloba* (L.) Taub. *Cytol.* **42** : 147-156.

Satyaprakash, T. L. and Singh, J. V., 2000. Evaluation and characterization of guar germplasm. M.Sc. Thesis, CCSHAU, Hisar.

Sen, N. K. and Vidyabhushan, R. V., 1960. Studies on tetraploid clusterbean varieties and their triploid and aneuploid progenies. *Cytol.* **25** : 426-436..

Sindhan, G. S. and Prashar, R. D., 1981. Correlation studies on bacterial blight incidence in guar germplasm with mineral and phenolic contents of leaves. *Proc. Third Int. Symp. on Plant Pathology*, IARI, New Delhi, Dec. 14-18, pp. 192.

Sindhu, A. S., Pandita, M. L., Arora, S. K. and Vashistha, R. N., 1982. Studies on genetic variability and correlations in clusterbean [*Cyamopsis tetragonoloba* (L.) Taub.]. *Haryana Agric. Univ. J. Res.* **12** (2) : 225-230.

Singh, A., 1972. Trisomics in *Cyamopsis psoralioides*. *Canad. J. Gen. Cytol.* **14** : 200-204.

Singh, A. K. and Sharma, S. K., 1965. Durgapura safed – a new guar for Rajasthan. *Indian Fmg.* **14** (10) : 23.

Singh, H. B. and Sikka, S. M., 1955. If it is guar here are two new Pusa strains. *Indian Fmg.* **5** (1) : 3-11.

Singh, J. V., Arora, S. K., Saini, M. L., Singh, V. P. and Gandhi, S.K., 1989. Genetic components of variance for bacterial blight and biochemical parameters in clusterbean. *Forage Res.* **15** (2) : 167-170.

Singh, J. V., Arora, S. K., Saini, M. L., Singh, V. P. and Gandhi, S. K., 1989. Biochemical differences in bacterial blight resistance in resistant and susceptible cultivars of guar. *Guar Res. Ann.* **5** : 1-3.

Singh, J. V., Lodhi, G. P. and Saini, M. L., 1990. Inheritance of leaf hairiness and leaf margin in clusterbean [*Cyamopsis tetragonoloba* (L.) Taub.]. *Ann. Arid Zone* **19** (1) : 55-57.

Singh, J. V., Lodhi, G. P. and Singh, V. P., 1995. Varietal improvement in guar. *Farm Dig.* **5** (1) : 23.

Singh, J. V., Lodhi, G. P., Saini, M. L. and Singh, V. P., 1990. Inheritance of branching and pods clustering pattern in clusterbean [*Cyamopsis tetragonoloba* (L.) Taub.]. *Crop Res.* **3** (1) : 93-95.

Singh, J. V., Saini, M. L., Lodhi, G. P. and Arora, S. K., 1990. Genetics of gum content in guar [*Cyamopsis tetragonoloba* (L.) Taub.]. *Forage Res.* **16** (1) :42-45.

Singh, J. V., Saini, M. L., Lodhi, G. P. and Arora, R. N., 1997. Estimation of catalase activity in healthy and bacterial blight infected leaves in segregating generations of clusterbean. *Third Agri. Sci. Cong.*, held at PAU, Ludhiana.

Singh, J. V., Saini, M. L., Arora, S. K., Singh, V. P. and Gandhi, S. K., 1990. Relationship between bacterial blight severity and biochemical parameters in clusterbean. *Ann. Biol.* **6** (2) : 117-120.

Singh, J. V., Saini, M. L., Singh, V. P. and Lodhi, G. P., 1990. Genetics of components of seed yield in clusterbean. *Guar Res. Ann.* **6** : 1-5.

Singh, J. V., Saini, M. L., Singh, V. P., Arora, S. K. and Lodhi, G.P., 1991. Gene effects for protein content in clusterbean. *Crop Res.* **4** (2) : 284-287.

Singh, J. V., Saini, M. L., Lodhi, G. P. and Arora, R. N., 1997. Genetics of bacterial blight resistance in clusterbean. *Ann. Arid Zone* **36** (20) : 143-146.

Singh, V. P., Yadav, R. K. and Chowdhury, R. K., 1981. Note on a determinate mutant of clusterbean. *Indian J. Agri. Sci.* **51** (9) : 682-683.

Sahoo, M. S. and Gill, K. S., 1976. Colchicine induced autotetraploidy in five genotypes of clusterbean [*Cyamopsis tetragonoloba* L. Taub.]. *Crop Imp.* **3** : 101-108.

Sahoo, M. S. and Gill, K. S., 1976. Inheritance of seven qualitative characters in clusterbean [*Cyamopsis tetragonoloba* (L.) Taub.]. *Genet. Agrar.* **30** : 185-190.

Srivastava, L. M. and Naithani, S. P., 1964. Cytogenetical studies in certain minor pulses and beans. *Cytol.* **29** : 453-464.

Swamy, L. N. and Hashim, M., 1980. Experimental mutagenesis in guar : Some induced viable mutations of systematic interest. *J. Cytol. Genet.* **15** : 61-63.

Tikka, S. B. S., 1975. Interrelationship between yield and components in clusterbean. *Indian J. Genet.* **35** : 340-343.

Tikka, S. B. S., Jaimini, S. N. and Sachan, S. C. P., 1974. Genotypic and phenotypic variability in collection of clusterbean varieties under high fertility conditions. *Indian J. Agri. Res.* **9** : 153-154.

Tripathi, R. M. and Lal, S., 1975. Estimates of variability and heritability of some quantitative traits in guar [*Cyamopsis tetragonoloba* (L.) Taub.]. *Indian J. Farm Sci.* **3** : 28-31.

Tyagi, I. B., 1974. Artificial medium for the germination of guar [*C. tetragonoloba* (L.) Taub.] pollen. *Pl. Sci.* **6** : 21-23.

Vig, B. K., 1963a. A note on the chromosome structure of guar (*Cyamopsis psoraliosoides* DC). *Sci. and Cult.* **29** : 95-96.

Vig, B. K., 1963b. Frequency of trivalents in autotetraploid guar. *Sci. and Cult.* **29** : 375-376.

Vig, B. K., 1965. Effect of reciprocal translocation on cytomorphology of guar. *Sci. and Cult.* **31** : 532-533.

Vig, B. K., 1969. Studies with Co^{60} radiated guar [*Cyamopsis tetragonoloba* (L.) Taub.]. *Ohio J. Sci.* **69** : 18.

Vijay, O. P., 1988. Genetic variability, correlations and path analysis in clusterbean [*Cyamopsis tetragonoloba* (L.) Taub.]. *Indian J. Hort.* **45** (1-2) : 127-131.

Whistler and Hymowitz, 1979. Guar: Agronomy, Production, Industrial Use and Nutrition. Purdue University Press West Lafayette, Indiana, pp. 124.

CHAPTER - 3

AGRONOMY

B.D.Yadav, Virendar Kumar and Anurag Saxena

Clusterbean [*Cyamopsis tetragonoloba* (L.) Taub.] commonly known as *guar* is an important leguminous crop of *kharif* grown in arid and semi-arid regions of North-Western India. Being drought hardy, the crop is being cultivated under rainfed conditions. It is grown as feed, fodder, green manuring, vegetable and seed, mainly in the states of Rajasthan, Haryana, Gujarat, U.P. and Punjab. Being legume, it builds up soil fertility and, thus, has a great role to play in nitrogen economy for the succeeding crop. With its diversified uses, it has now become an important industrial crop with a great potential for foreign exchange. Industrial importance of the crop has given a new dimension to its cultivation and created an interest among the research workers, growers and industrialists. Research work done in India on some of the important agronomic practices influencing yield and quality of clusterbean grown for grain, is reviewed, here.

SOWING VARIABLES

Sowing time

The sowing time is one of the important variables bringing about dramatic changes in growth environment of the crop, and consequently the yield variables. There are numerous reports which indicate that yield is considerably reduced when sowing is delayed to last week of July or first week of August.

Similarly, early-sown clusterbean makes luxuriant growth and the crop is damaged due to lodging. Bains and Dhillon (1977) conducted experiments on various sowing dates of clusterbean at Ludhiana and observed that first week of July was the best sowing-period compared to third week of July and first week of August, giving 19.4 and 73.6 per cent higher seed-yield over the later respective dates. In a study at Hisar, Sharma *et al.,* (1984) observed that maximum seed yield of guar could be realized with 5 July sowings in comparison to that sown on 20 June, 20 July and 5 August. In another trial conducted at Hisar, 15 July was found optimum sowing time. Too early (1 June, 15 June and 30 June) and too late (30 July) sowings resulted in lower seed yield as compared to 15-July sowing, for instance, respective reductions in yield were to the magnitude of 41.77, 19.73, 6.80 and 38.40 per cent (Singh *et al.,* 1979). Similar results on seed yield were also reported by Das *et al.,* (1977), Singh *et al.,* (1980) and Tomer and Singh (1976). Two year study conducted at Jhansi,revealed that sowing done between 15 and 25 July, gave significantly higher grain yield than early (25 June – 5 July) or late (5 August) sowings. It was therefore, concluded that for higher seed yield on drylands of Bundelkhand region (UP), clusterbean could be sown on 15 to 25 July (Ali *et al.,* 1978). At Jodhpur, the seed yield of clusterbean declined with delayed in sowing beyond 1 July, and the reduction was to the extent of 32.5 and 63.7 per cent, when sowing was delayed upto 30 July and 16 August, respectively (Singh, 1977a).

Based on work conducted at Hisar, Jain *et al.,* (1987) recorded maximum value of yield (16.75 q/ha), protein (5.65 q/ha) and gum yield (4.85 q/ha) at 10-July sowing. Similarly, Taneja *et al.,* (1995) reported 10 July as the optimum sowing time which gave maximum seed yield (21.7 q/ha), whereas further delay in sowing to 30July caused drastic reduction in seed yield (12.6 q/ha). Bhadoria and Chauhan (1994) also got the similar trends relating to sowing time on yield and quality. Dhukia and Singh (1988) indicated high crude-protein content, protein yield and gum yield with 10-July sowing. The gum content in seed, however, increased with delay in sowing. Dry-matter accumulation, leaf-area index and seed yield were higher with 10-July sowing in comparison to delayed sowings to 25 July and 9 August (Dhukia *et al.,* 1990).

Sowing Systems

In a preliminary study carried out at CAZRI, Jodhpur, it was revealed that *guar* yield in the paired row system of sowing was obtained by 8 and 46 per cent higher over the traditional system in wet and drought years, respectively (Singh, 1977a). Consistently higher seed yields over the uniform system of sowing were obtained when *guar* was sown in the paired row system, particularly in a drought year (Singh and Singh, 1975).

Spacings

Establishment of optimum plant popualtion per unit area of land, ensures proper utilisation of inputs *viz*., nutrients, moisture and light resulting in better performance of plant in the plant community. Cultivars of clusterbean are reported to show variation as regards their growth habit and yield potential (Mittal *et al.,* 1977) and respond differentially to plant population (Singh and Tomer 1975).

Inter row spacings (30, 45, 60, 90 cm) effect on the growth and yield of guar studied under rainfed condition at CAZRI, Jodhpur for two years (1975 and 1976) revealed that grain yield decreased with increasing row spacings, while yield per plant increased with increasing row spacings (Singh, 1977a). A combination of 45 inter and 15 to 20 cm intra row-spacings was found to be optimum for variety FS 277 when planted at normal sowing time (middle of July). For late sown crop (14 August), a row spacing of 30 cm, instead of 45 cm, was found optimum (Anonymous, 1976). Sharma *et al.,* (1984) and Taneja *et al.,* (1984) obtained higher seed yield at 30 cm row spacings for variety FS 277. Similar studies at Hisar indicated that intra-row spacings of 15 cm gave significantly higher grain yield (18.03 q/ha) and maximum gum content over 7.5 cm (15.97 q/ha) and 22.5 cm(15.0 q/ha) sowing (Das *et al.*, 1977).

In AICRP trials conducted during *kharif* 1977 at 5 locations (Jodhpur, Bikaner, Durgapura, Junagarh and Agra), it was observed that higher yield of variety FS 277, 2470 (2), Durgapur Safed and KVS-1 could be achieved with inter row spacings of 60, 45 and 30 cm, respectively (Singh 1978). At Hisar, 45 cm row spacing was found suitable for HG 75, Durgapur Safed (Taneja *et al.*, 1982, Yadav *et al.,* 1989) and HG 182 (Jain *et al.,* 1987), whereas 30 cm row spacing was optimum for FS 277 (Taneja *et al.,* 1982, Yadav *et al.,* 1989).

Bains and Dhillon (1977) recommended a spacing pattern of 30 x 22.5 and 45 x 15 cm at Ludhiana. However, 45 x 15 cm can be adopted safely. At Jhansi, row spacing of 60 cm (1.1 lakh plants/ ha) was optimum for Durgapur Safed, 45 cm (1.5 lakh plants/ ha) for B19-1-55 and 30 cm (2.22 lakh plants/ha) for FS 277 (Ali, 1982). However, Singh (1982) suggested a row spacing of 45 cm (1.25 to 1.50 lakh plants/ha) for optimum seed yield for branched variety HG 75 at Dryland Research Station, Bawal (Haryana).

Rana *et al.,* (1991) obtained higher seed yield with 30 cm row spacing than 45 cm under late sown conditions for variety HG 75, Naveen and PLG 85. An intra-row spacing of 15 cm was more suitable for branched and 10 cm for unbranched varieties. Higher LAI, LAD and CGR were observed with 30 cm row spacing, but seed yield was more with 45 cm row spacing (Yadav *et al.,* 1990). In further studies carried out at Hisar, Yadav *et al.,* (1991) found that total dry-matter, N and P content in seed and straw were not influenced by different row spacings. However, total uptake of N and P was significantly higher in 45 cm row spacing than 30 cm row spacing.

Seed Rate

In a study carried out at Hisar, seed rates of 20 and 30 kg being at par, were found superior to 10 kg/ha, but gum and protein content remained unaffected with variation in seed rate (Singh *et al.,*1978). Singh and Singh (1989) working at Bichpuri obtained higher seed yield with seed rate of 30 (19.19 q/ha) over 10 (14.1 q/ha) and 20 kg/ha (16.8 q/ha). Yield attributes, gum and protein content, decreased with increase in seed rate. Tiwana and Tiwana (1992) got 17.9 and 8.2 per cent higher seed yield with 30 kg over 20 and 40 kg seed/ha, respectively.

CLUSTERBEAN IN SEQUENTIAL SYSTEM

Amongst the crops preceding cotton, clusterbean had the most beneficial effect with regard to yield and quality (Negi and Singh,1955). In experiments carried out at Jodhpur, inclusion of legumes in rotation with pearl millet *viz.,* pearl millet-clusterbean and pearl millet-green gram increased the total productivity per unit area over pearl millet-pearl millet, pearl millet-sorghum and pearl millet-fallow systems of cropping (Misra, 1964). Clusterbean was identified as the most useful and remunerative crop preferably in wheat based cropping system (Reddy *et al.,* 1981). Saxena *et al.,* (1997) also obtained higher grain yield of pearl millet (23.6%) when it was grown after clusterbean, compared to continuous pearl millet.

When barley followed clusterbean or sun hemp, its grain

and straw yields were higher than that grown after fallow. However, when soil moisture was limiting, green manuring of these crops proved less beneficial to the succeeding barley, as compared to the same used for fodder (Singh and Gautam, 1973). Studies conducted at Sriganganagar, showed that amongst various *kharif* crops (cotton, green gram, pigeon pea, clusterbean) preceding wheat, clusterbean–wheat rotation recorded maximum water-use efficiency (Singh *et al.,* 1998). In the fifth successive year, good crop of clusterbean and wheat could be raised with saving of 26 kg P_2O_5/ha in clusterbean and 30 kg N/ha in wheat without sacrificing the yield of the two crops following clusterbean–wheat crop rotation. In other clusterbean based crop rotation, crop accompanying/following clusterbean continued to get benefit of N fixed by clusterbean to the extent of 30 kg N/ha at Gwalior and Delhi (Anonymous, 1996).

De *et al.,* (1983) observed that yield of subsequent rice was more with preceding short-duration legumes (green gram, black gram, cowpea) in comparison to long-duration legume *i.e.* clusterbean or maize. However, at Bawal, the yield of subsequent mustard was more with preceding clusterbean, and lower with subsequent pearl millet or cotton as compared to fallow-mustard sequence (Anonymous, 2000). The grain and straw yield of wheat were not influenced by residual effect of clusterbean, grown either for grain or fodder, but there was an addition of 60-80 kg N/ha in the soil, over the total nitrogen status, due to inclusion of clusterbean in the sequence (Regmi *et al.,* 1984). The uptake patterns of nitrogen and phosphorus by wheat grain were similar, where clusterbean was taken for grain or fodder, but there was higher uptake of nitrogen by straw, when clusterbean was grown for fodder than for grain (Regmi *et al.,* 1985).

CLUSTERBEAN IN INTERCROPPING SYSTEM

This practice improves soil fertility and also has the yield-advantages to the companion cereal crop (Singh and Ahuja, 1990), which may contribute to net returns also. The other logical point professing for intercropping, is to provide insurance against complete crop-failure under any biotic or abiotic interferences.

Inter/mixed/strip cropping with cereals

Earlier studies indicated that mixed cropping of clusterbean was better than its intercropping with pearl millet or sole cropping of the later (Rana *et al.*, 1986), wherein clusterbean yielded 6.8 q/ha by mixed cropping and 3.3 q/ha in intercropping system without any reduction in yield of pearl millet. Keshwa and Singh (1991) reported higher biomass production and green-fodder yield of pearl millet mixed with clusterbean than sole pearl millet. Mixture of clusterbean in sorghum, increased forage yield (Rao *et al.*, 1979) and improved its quality.

In general, intercropping of clusterbean with pearl millet has been found profitable in good and sub-normal rainfall years, but unprofitable in drought years (Daulay *et al.*, 1998). Studies carried out at Hisar (Yadav and Joon, 1993) and Gwalior (Bhadoria *et al.*, 1992), on intercropping of pearl millet with clusterbean revealed that higher returns were obtained with the sole crop of clusterbean, implying that intercropping with pearl millet in any proportion had no economic significance. However, amongst different intercropping systems, 3:1 ratio of clusterbean : pearl millet gave highest clusterbean equivalent and economic return closely followed by 2:1, indicating that where chances of clusterbean failure are high due to uncertain weather, it may be intercropped in 3:1 ratio to divide risk against complete failure. At Jodhpur, evaluation of performance of various clusterbean genotypes as inter/mix crop in pearl millet, showed that mixed cropping was more beneficial in good rainfall years than drought years and vice-versa. Variety FS 277 was most suitable and proved beneficial for inter-sowing under good rainfall, sub-normal rainfall or drought condition except variety 2470/12 which proved better for mixed cropping in good rainfall year only (Daulay *et al.*, 1998).

Out of three legumes as inter crop, clusterbean proved significantly superior to cowpea and green gram at Jobner and gave 2.95 q/ha additional grain yield when two rows of clusterbean were sown in between paired rows of pearl millet with application of 30 kg N/ha (Bhati and Manohar, 1989). At Bawal, two rows of clusterbean sown in inter-spaces of pearl millet planting system of 100 cm apart, gave higher total productivity and gross returns as compared to sole crop of pearl millet and proved better than even sole crop of clusterbean (Singh, 1981). Consumptive use of water, water-use efficiency, land-equivalent ratio and productivity were higher in clusterbean intercropping than sole pearl millet, both under rainfed and protective-irrigation condition (Singh, 1985). In a 2-year study on arid land, double row of clusterbean planted in interspaces of paired rows of pearl millet (30/70 cm), yielded

3.81 q/ha additional grain without affecting the pearl millet yield and gave an advantage of 21% in land-equivalent ratio (Muthuvel *et al.,* 1984). Intercropping of 1 or 2 rows of clusterbean earned higher net returns, without causing yield reduction in the base crop of pearl millet (Sharma *et al.,* 1988) at Jobner. It increased growth, yield attributes and yield of grain and straw of pearl millet (Singh *et al.,* 1989) in similar intercropping system at Jobner. At Bikaner, strip-cropping of pearl millet : clusterbean in 4:4 row ratio gave highest pearl millet yield than 4:6 or 4:8 ratio (Yadav, 1992). Whereas, Singh and Joshi (1994) observed the superiority of alternate row planting (1:1) of pearl millet : clusterbean in comparison to other systems of planting. Although clusterbean yield was highest in 4:4 row strip-cropping of pearl millet : clusterbean, yet the net returns were highest in alternate-row intercropping system (Singh and Joshi 1994a). At Jodhpur higher total productivity and water-use efficiency were obtained with 1:1 row intercropping than 4:4 strip-cropping or sole pearl millet (Singh and Joshi 1997). Evaluation of different pearl millet based cropping systems in drought amelioration revealed that strip-cropping in companionship of clusterbean, gave better yields and covered more risk compared to other systems (Singh and Joshi, 1994a).

Crude-protein in sorghum increased with clusterbean intercropping (Sharma *et al.,* 1993). Beneficial effects of clusterbean intercropping were also observed on grain yields of maize (Singh and Kaushik, 1997) and sorghum (Singh and Ahuja, 1990). Intercropping one row of clusterbean between the rows of rainfed maize increased the total grain yield, maize-grain protein content and net profit, compared with sole crop of maize (Chouhan and Dungarwal, 1982).

Intercropping with oilseeds

At Bawal, intercropping 2 rows of clusterbean in castor grown in paired rows (30/120 cm), increased the castor-equivalent yield, land-equivalent ratio and net returns compared to that castor in pure stands (Singh and Singh, 1988). Castor yield declined with clusterbean intercropping, whereas yield reduction in clusterbean due to companionship of castor was to the extent of 50 per cent (Venkateswarlu and Subramanian, 1990). Studies at CRIDA, Hyderabad revealed that sowing of castor and clusterbean in companionship, reduced their respective yield by 20 and 70 per cent, respectively, compared to their pure stands (Reddy and Venkateswarlu, 1989), indicating higher sensitivity of clusterbean in mixed stand.

Clusterbean, as an intercrop in sunflower, gave highest yield compared to other grain legumes, but adversely affected the sunflower yield at Jodhpur (Singh and Singh, 1977). However, sunflower as an intercrop reduced the stem height, leaf area and grain yield of all companion crops including clusterbean (Narwal and Malik 1986).

Intercropping groundnut with clusterbean in 150 cm apart rows, provided an additional bonus yield (2.0 q/ha) of clusterbean with highest returns (Dhillon *et al.*, 1980), whereas, intercropping of sesame, in alternate rows or in between the paired rows of clusterbean, caused reduction in its yield, but earned maximum returns due to higher market-price of the intercrop's produce (Singh, 1981c).

Intercropping with pulses

Under irrigated condition at Hisar, clusterbean intercropping reduced the grain weight, pods/plant, branches/ plant and dry-matter accumulation, but increased the plant height of the base-crop pigeon pea. Clusterbean, as an inter- crop, yielded 7.0 q/ha of grain, but caused drastic reduction (9.7 q/ha) in pigeon pea yield, and thus, proved a losing proposition (Kumar and Agarwal 1982). At Morena, Rajput *et al.*, (1989) observed higher pigeon pea equivalent yield with clusterbean intercropping (25.9 q/ha) compared to sole pigeon pea (20.8 q/ha), but Yadav *et al.*, (1997b) found that amongst all intercrops, clusterbean yielded lowest and caused maximum yield reduction in base crop pigeon pea, thus proved most uneconomical. However, no such effect of clusterbean as an intercrop, in increasing pigeon pea equivalent yield, was noticed by Rai and Sinha (1993).

Intercropping of urd bean in clusterbean, at Bawal, showed that none of the planting systems could out-yield the total productivity of sole clusterbean sown at 30 cm row-spacing (Singh, 1981a).

Intercropping with grasses

Work on intercropping system for perennial grasses in dryland at CAZRI, Jodhpur has shown that in good rainfall (500 mm) years, clusterbean proved to be a good companion crop to grasses, which resulted in increased total productivity, land utilization (1.73) and moisture-use efficiency (Daulay 1977). In initial year of grass-planting at Bawal, maximum yield (7.5 q/ha) of intercrop (clusterbean) was obtained when planted in 1:2 grass-legume ratio, but it was at the cost of green-fodder yield of grass. In the succeeding year, no yield of clusterbean was, however, obtained due to seedling mortality, as the rhizosphere was soon rendered siccative by the already established grass cover (Kumar and Agarwal 2000). Cultivation of *cenchrus ciliaris* in mixture with clusterbean, for fodder production, increased the crude protein and *in vitro* digestibility as compared to sole *Cenchrus ciliaris* (Gill *et al.*, 1984).

Intercropping with sugarcane

Intercropping clusterbean in sugarcane was most profitable, wherein the sugarcane yield (100.8 t/ha) was comparable to that obtained from sole sugarcane (101.2 t/ha) cropping system (Deshpande and Nanker 1977). Sugarcane + clusterbean intercropping was found to give higher net profit than sole cropping of sugarcane (Dixit and Misra, 1991).

Intercropping with tree species

Under agri-silviculture system, clusterbean yield was adversely affected by *subabul* [*Leucaena leucocephala* (Lamk) de Wit] with advancement in its age and canopy development. The content and uptake of N in seed and stalk of clusterbean was maximum in the 1.5 m topping-system and comparatively more in the first year, because of less competition offered by *subabul* for light, moisture and minerals. Clusterbean gave 3.0 q ha^{-1} yield when grown in association with established (5-year old) *ber* orchard after light pruning under rainfed condition at Bawal (Kumar *et al.,* 1999).

Intercropping with other crops

Studies at Banglore on irrigated land, recognised the feasibility to intercrop medicinal yam with clusterbean without affecting its tuber yield and N requirement. Though, this combination had a higher yield potential, it gave lower monetary profit and N-use efficiency, than other crop-combinations with medicinal yam (Hegde and Randhawa, 1981). Vetiver (*Vetiveria zizaonioides*) + clusterbean inter-cropping gave highest overall returns followed by vetiver + blackgram.

FERTILIZER MANAGEMENT

Effect of Nitrogen

Though leguminous crops generally, do not respond to nitrogen, yet a small dose may stimulate growth of the plants in early stages. Atwal and Sidhu (1964) indicated that application of N in small doses improved nodulation, growth and nitrogen fixation in clusterbean. At Agra, higher grain and straw yield, nodules/plant, root dry-weight, water-use efficiency, gum content and net return were found with 20 kg N/ha over no-nitrogen (Singh and Singh,1989). However, protein content remained unaffected with N application. Similar effect of N on grain yield, gum content and protein content were observed elsewhere. On sandy-loam soil at Hisar, application of 20 kg N/ha led to significant increase in physiological parameters *viz.*, crop-growth rate, leaf-area index and leaf-area duration (Yadav *et al.*, 1990), yield attributes, seed and stalk yield (Yadav *et al.*, 1989), dry-matter accumulation and uptake of N and P in seed (Yadav *et al.*, 1991), and yield of protein and gum (Yadav *et al.*, 1989a). However, Gill and Singh (1981) found no response of 20 kg N/ha on yield attributes, grain yield and stalk yield of clusterbean. Nitrogen application was found to increase N, P, K contents and their uptake in seed and straw (Singh and Singh, 1990). Studies carried out at Hisar showed that 30 kg N/ha increased the seed yield, N-uptake and concentration of N in seed and straw (Singh *et al.*, 1993).

Effect of phosphorus

P application may benefit nodulation in clusterbean (Misra, 1961). Higher net-assimilation rate (Raut and Ali 1983), crop growth rate, leaf-area index, leaf-area duration (Yadav *et al.*, 1990) and root dry-weight (Singh and Singh, 1989) were related to phosphorus application. Numerous reports are available on the phosphorus application in improvement of growth and yield of clusterbean (Singh and Singh 1989).

A study on cultivators' field in Haryana revealed that P alone and in combination with N, increased the seed yield significantly over control, whereas, the response in kg/kg of applied P was 9.0 and 8.2 at 20 and 40 kg P_2O_5/ha, respectively (Dahiya *et al.*, 1986a). Seed yield of clusterbean increased upto 40 kg P_2O_5/ha and there was a noticeable fall in seed yield with 60 kg P_2O_5/ha (Bhardwaj *et al.*, 1982).

In some studies at Hisar, a dose of 40 kg P_2O_5/ha was found superior to control in terms of seed and stalk yield (Yadav *et al.,* 1991), dry-matter accumulation, crop-growth rate, leaf-area index, leaf-area duration (Yadav *et al.,* 1990), yield-attributing characters *viz.*, number of clusters/pod, number of pods/plant and seed yield/plant (Yadav *et al.,* 1989). However, in some studies, there was no response of phosphorus beyond 20 kg P_2O_5/ha (Taneja *et al.*, 1981) or 30 kg P_2O_5/ha (Gill and Singh 1981) under rainfed condition. On loamy-sand soil of Bawal with medium P status a dose of 30 kg P_2O_5/ha was found as the economic dose (Dahiya *et al.*, 1986).

There was a steady increase in yield with increasing levels of phosphorus upto 60 kg P_2O_5/ha, but 40 kg P_2O_5/ha was found to be optimum (Menon 1977). At Durgapura, the response of phosphorus was noticed up to 20 kg P_2O_5/ha. Work carried out at CAZRI, Jodhpur showed that clusterbean responded up to 40 kg P_2O_5/ha, and the increase in yield beyond this level was marginal (Singh, 1977a).

On loamy-sand soil at Jobner, application of 40 kg P_2O_5/ha was found superior to 20 kg with regard to increase in growth, yield attributes, seed yield and stover yield (Shivran *et al.,* 1996), whereas, in earlier study at the same location, Meena *et al.,* (1991) reported 30 kg P_2O_5/ha as a optimum dose. Similar response of phosphorus was observed at Bhatinda (Singh and Tiwana 1995). Recent work at Bikaner revealed significantly higher response on yield of seed and straw with the application of 40 kg P_2O_5/ha over 20 kg and control (Solanki *et al.,* 1998).

At Agra, there was a linear response of applied P on sandy-loam soil, wherein 60 kg P_2O_5/ha was superior to 30 kg and control in terms of grain yield, monetary returns and water-use efficiency (Singh and Singh 1989). At Gwalior, clusterbean yield increased with each successive increase of 20 kg P_2O_5/ha, up to a level of 60 kg P_2O_5/ha and increase in yield with 60 kg P_2O_5/ha was 44.9 per cent over control (Bhadoria *et al.,* 1997). Khan and Guriar (1997) also found 60 kg P_2O_5/ha as the suitable dose for clusterbean production at Gwalior.

Singh and Singh (1989) observed a significant increase in the content and yield of protein and gum with the application of 60 kg P_2O_5/ha over 30 kg P_2O_5/ha and control. Similar findings were observed in the studies carried out at Gwalior (Bhadoria *et al.,* 1997). Yadav *et al.,* (1989a) observed that content of protein and gum remained uninfluenced with phosphorus application, but yield of protein and gum increased with 40 kg P_2O_5/ha over control. Similar effects of phosphorus on quality were observed by Gill (1979). Jain *et al.,* (1988) revealed that gum content increased with the application of phosphorus @ 20 kg P_2O_5/ha (29.41%) over control (27.95%). Higher doses of phosphorus (40 and 60 kg P_2O_5/ha) resulted in lowering of gum content, though it was higher (29.04 and 28.87%) than the control value. The extent of increase in protein content was recorded as 1.17, 2.50 and 2.81 per cent over control, with 20, 40 and 60 kg P_2O_5/ha, respectively.

Nodulation in clusterbean was found positively correlated to phosphorus uptake leading to higher nitrogen fixation (Rewari *et al.,* 1965). Gill (1979) observed that P content and its uptake improved in plant by phosphorus fertilization. Yadav *et al.,* (1991) found significantly higher uptake of N and P in both seed and stalk of clusterbean, with the application of 40 kg P_2O_5/ha, over control. Singh and Singh (1990) also found increase in both contents and uptake of N, P and K with increasing rates of P_2O_5 (0, 30, 60 kg/ha).

Effect of sulphur

Legumes respond to sulphur application in sandy soils of North-Western India owing to be deficient in available S. Screen-house study at Hisar showed that in clusterbean stem and leaves, the concentration and uptake of N increased significantly with the application of 60 mg S kg^{-1}, while of P increased only up to 30 mg S kg^{-1}. The increase in concentration and uptake of K by application of S was recorded upto 60 mg S kg^{-1} in leaves and up to 90 mg S kg^{-1} in stem. Application of S increased the concentration and uptake of S in leaves and stem up to 90 mg S kg^{-1}.

Sulphur is known to increase growth, yield attributes, straw yield, seed yield and harvest index up to a dose of 60 kg S/ha on loamy-sand, S-deficient soil under rainfed condition at Jobner in Rajasthan (Shivran *et al.,* 1996). However, clusterbean grown on S-deficient soil of Gwalior, was found to respond up to a level of 40 kg S/ha in terms of growth, yield attributes, grain yield as well as quality characters *viz*., phosphous, sulphur, gum and protein content in grain (Bhadoria *et al.,* 1997). Similar results were obtained by Khan and Guriar (1997) at Gwalior. Based on 3 years' study (1992-1994), application of 20 kg S/ha resulted in significantly higher seed yield (12.12 q/ha) and gum content (29.98%) over control (10.82 q/ha, 29.10%) at Bawal (Kumar *et al.*, 1999). At Durgapura, the crop was found to respond up to 20 kg S/ha, (Table 1) whereas, no response of S was noticed at S.K.Nagar in Gujarat.

Table 1. Effects of S on seed yield and cost-benefit ratio

S levels (kg/ha)	Seed yield (q/ha)			Cost-Benefit ratio		
	Bawal	Durgapur	Gwalior	Bawal	Durgapur	Gwalior
0	10.82	7.25	15.81	–	–	–
20	12.12	8.42	18.40	7.09	10.89	12.23
40	11.95	8.10	20.29	3.11	4.00	10.58
CD at 5%	0.95	0.58	0.89	–	–	–

Studies carried out on the effect of varying S-sources on S-deficient sandy-loam soil showed that gypsum was better than pyrite.The number of seeds/pod and branches were higher with gypsum than pyrite (Shekhawat *et al.*, 1996).

Effect of micro-nutrients

Increase in irrigation facilities due to intensive cropping and lesser use of organic manure have rendered 50 per cent of the Indian soils deficient in Zn (Katyal and Sharma, 1979). In earlier studies at Ludhiana, soil application of 50 kg $ZnSO_4$/ha was found to increase clusterbean yield under acute Zn-deficiency. Zn applied to soil, was more effective in increasing yield than that applied as foliar spray (Takkar *et al.*, 1973). At Jobner, application of zinc @ 10 kg/ha produced significantly higher primary branches, pods/plant and grain yield over no zinc (Maliwal *et al.*, 1987). At Hisar, 20 kg $ZnSO_4$/ha was found to cause significant increase in seed yield and gum content over control (Table 2).

Table 2. Effect of $ZnSO_4$ on seed yield and quality of cluster-bean seed (Location, Hisar)

$ZnSo_4$ levels (kg/ha)	Seed yield (q/ha)	Gum (%)	Protein (%)	Gum yield (q/ha)
0	13.19	29.25	28.73	3.86
10	14.20	30.77	29.06	4.37
20	15.10	31.13	28.37	4.70
CD at 5%	0.82	0.46	0.32	0.38

Studies carried out at Durgapura showed that foliar spray of 1 kg Zn/ha (0.2%) at 45 days of sowing was better than soil application of 5 or 10 kg Zn/ha. Pot-culture studies on the effect of source ($ZnSO_4$, ZnO), dose (5 and 10 kg/ha) and mode (soil application, seed coating, foliar spray) of Zn application revealed that 3 sprays of 0.5% $ZnSO_4$ at 15-day interval proved more effective, which increased protein as well as carbohydrate content of both leaves and nodules, and significantly improved the number of pods, seed weight, leaf area, nitrogen content and nitrogenase activity per plant (Nandwal *et al.*, 1990). Screen-house study carried out at Hisar, on the effect of micro-nutrients *viz.*, Zn, Cu, Mn and Fe, applied individually or collectively, revealed that Zn alone or in combination with other nutrients, significantly increased the dry-matter yield, its concentration and uptake. Gupta (1983) conducted studies at Hisar on comparative response of some *kharif* crops to Zn application and observed that pulse crops i.e. black gram and green gram, clusterbean and soybean responded to Zn application up to 5 ppm. Soybean was more resistant than clusterbean, black gram and green gram, and the response was 69 per cent against 210, 611 and 1014 per cent in case of clusterbean, black gram and green gram, respectively. The deficiency was mitigated when 5 ppm Zn was applied. Singh *et al.*, (1989b), while working at Hisar, reported that among different zinc sources, zinc sulphate showed the best results followed by zinc phosphate.

Irrigation management

Studies carried out by Meena *et al.*, (1991) on loamy-sand soil at Jobner (Rajasthan) indicated that one life-saving irrigation at 60 days after sowing, increased the grain yield by 32.1 per cent over no-irrigation (Table 3). Earlier study on clusterbean, grown in summer, seed yield was significantly reduced in case number of irrigation was reduced from three to one (Gill *et al.*, 1974). At Sriganganagar (Raj.), grain yield increased by 16.7 and 27.3%, when the crop was given 1 and 2 irrigation, respectively, over control (Singh *et al.*, 1998). One post-sowing irrigation scheduled at 275 mm CPE, resulted in significant increase in seed yield, dry-matter accumulation and leaf-area index at 80 days of sowing and at harvest (Dhukia *et al.*, 1990). The seed protein content was significantly improved when the crop was irrigated at 55 days of sowing, but irrigation beyond third week of September deteriorated the seed quality (Singh *et al.*, 1980).

Table 3. Effect of irrigation on seed yield and dry-matter accumulation

Irrigation levels	Seed yield (q/ha)		Dry matter/plant (g)	
	Jobner	Hisar	Jobner	Hisar
No irrigation	6.64	14.12	38.00	48.92
One irrigation	8.77	16.10	43.47	58.08
CD at 5%	0.78	0.31	2.44	1.68

Source: Meena et al., 1991

WEED MANAGEMENT

Clusterbean is being grown during rainy season, hence, a large number of weeds may compete for nutrients, moisture and space causing considerable yield reductions (Daulay and Singh, 1982). Season-long competition with weeds in clusterbean causes severe yield-reduction ranging from 29-48 per cent (Bhadoria *et al., 1*996), and the severity may be higher (70-98%) depending on the weed infestation (Sonani et al., 1985). Reports indicate that in the absence of any other input, weed control alone is responsible to increase the seed yield by 68% (Yadav *et al.,* 1993) at Bawal and 61% (Yadav *et al.,* 1991a) at Hisar. The pre-dominant weed-flora that rock the growth and yield of the crop, vary with soil type, moisture condition and other climatic factors. The most common weeds that infest the clusterbean growing tract in India are *Cyperus rotundus, Trianthema portulacastrum, Digera arvensis, Tribulus terrestris, Amaranthus viridis, Cynodon dactylon, Celosia argentina, Phylanthus niruri, Echinochloa colonum, Pulicaria wightiana, Portulaca oleracea, Digitarea sanguinalis* etc (Kumar *et al.,* 1996, Bhadoria *et al.,* 1996), though the pre-dominance of one over other may vary from region to region.

Removing weeds at any time during the growing season may not be beneficial, but stage of weed control is more important (Table 4). Studies have shown that critical period of crop-weed competition lies from 15 to 30 days of sowing on loamy-sand (Kumar *et al.,* 1996) and sandy-loam (Yadav *et al.,* 1997a) soils. However, Sonani *et al.,* (1985) reported period between 20-50 days after sowing most critical. Removing weeds at 20 or 30 days increased the water-use efficiency and seed yield. Water-use efficiency decreased with increase in time of weed removal beyond 20 days after crop sowing (Yadav 1998). Since the crop is grown rainfed in wide rows, mechanical methods of weed control is more suited, as they provide additional benefit of conserving soil moisture. Obviously, the earlier research on weed control, had mainly been concentrated on mechanical methods. Studies conducted in different agro-climatic zones and on variant soil types, have established the superiority of mechanical/cultural means over chemicals (Yadav *et al.*,1990a, Kumar *et al.*, 1996 and Bhadoria *et al.*,1996). These methods, besides being environmental friendly, are comparatively more economic than chemicals (Yadav *et al.*, 1997, Bhadoria *et al.*, 1996). However, a need was felt to devise alternate methods of weed control where mechanical weed-control could not be adopted due to some inevitable circumstances *viz.*, unavailability of labour in peak season and unceasing rains during critical period (Table 5).

Table 4. Effect of crop-weed competition on seed yield and weed control efficiency

Treatments	Seed yield (q/ha)		Weed control efficiency (%)	
	Bawal	Hisar	Bawal	Hisar
Weedy upto				
15 DAS	9.65	13.7	85.2	82.4
30 DAS	8.82	12.6	70.0	70.2
45 DAS	7.06	10.2	50.3	29.6
60 DAS	5.98	7.9	21.7	8.5
Harvest	5.90	7.1	0.0	0.0
Weed free upto				
15 DAS	9.93	10.2	81.6	45.1
30 DAS	10.39	14.4	85.5	80.3
40 DAS	10.84	15.0	90.8	87.7
60 DAS	11.56	15.2	95.1	95.7
Harvest	11.64	15.4	100.0	100.0
CD at 5%	1.41	1.90	-	-

Table 5. Effect of weed control methods on seed yield and economics of clusterbean

Treatments	Seed yield (q/ha)			Net profit (Rs./ha)		
	Bawal	Hisar	Gwalior	Bawal	Hisar	Gwalior
Weedy check	9.9	9.5	9.5	781	115	4404
One hoeing	13.9	16.2	13.9	2225	2303	6909
Two hoeing	14.4	18.9	–	1684	2900	–
Weed free	14.8	19.9	15.7	1612	2768	7475
Fluchloralin 1.0 kg a.i./ha	13.8	–	–	2098	–	–
Fluchloralin 1.5 kg a.i./ha	14.2	18.4	14.42	1954	2845	6849
Pendimethalin 1.0 kg a.i./ha	10.3	11.6	13.63	1565	125	6467
CD at 5%	2.3	2.1	–	–	–	–

Preliminary evaluation of herbicides for forage cropping system at IGFRI, Jhansi, indicated that, of several herbicide tested, pre-emergence application of 1-3 kg bentazone and 0.5-1.5 kg benefin (benfluralin)/ha was effective against weeds of clusterbean (Dutta *et al.,* 1974). However, DSMA @ 2 kg/ha was found safe for post emergence use in clusterbean (Dutta *et al.,* 1975). Earlier studies revealed that Treflan, incorporated into top 5-cm soil prior to planting, is effective in controlling grasses and broad-leaved weeds. Use of TCA and Dalapon @ 6.5 kg a.i./ha reduced the frequency of *Cynodon dactylon,* but the same failed to control *Trianthema portulacastrum*, which comprised more than 50 per cent of the weed cover. Vernolate (PPTC) and EPTC proved very effective in controlling *T. portulacastrum* and broad-leaved weeds, respectively. Highest fresh-weight yield of forage was obtained with Vernolate at 4.5 kg a.i./ha. Work at Jodhpur showed that alachlor, trifluralin and nitrofen were effective in controlling weeds that gave yield at par to hand weeding. Alachlor was most effective chemical for weed control in clusterbean, whereas nitrofen was specifically effective on *Pulicaria wightiana* (Daulay and Singh 1982).

Recent work on chemical weed-control revealed that Basalin is an effective herbicide for controlling weeds, when incorporated in soil @ 1.0 kg a.i./ha on loamy sand (Kumar *et al.,* 1996) and @ 1.5 kg a.i./ha on sandy loam (Yadav *et al.,* 1990a, Yadav *et al.,* 1997, Bhadoria *et al.,* 1996) type of soil. Pendimethalin caused phytotoxic effect in clusterbean on loamy-sand soil (Kumar *et al.,* 1994), but no such effect with any of the three dinitroaniline herbicides tried, was observed on growth, yield (Budhiraja *et al.,* 1998) and quality (Budhiraja *et al.,* 1998a) of 4 clusterbean varieties tested on sandy-loam soil. This behaviour needs further investigation, taking into consideration various factors *viz.,* soil type, moisture, method and time of application, environmental condition and interaction of these herbicides on disease incidence in this crop.

FUTURE THRUST

It is quite explicit from the foregoing review that clusterbean research has gained momentum in the last 2 decades and lot of information has been generated on agronomic aspects. However, there is need to intensify research on those aspects that remained inconclusive, and to initiate research on those areas which lacked information. Considering this, the future line of work needs to be directed as below :

(i) Developing need based agro-techniques for higher productivity of the crop.

(ii) Developing production technology for newly evolved genotypes enabling them to fit into common crop rotations, especially the clusterbean-mustard.

(iii) Chemical weed control with special refernece to pre and post emergence weedicides.

(iv) Role of clusterbean in the nitrogen economy of succeeding crops in clusterbean based cropping system.

(v) Evaluation and development of drought amelioration, moisture conservation and water harvesting techniques/ practices for increasing productivity of clusterbean.

(vi) Optimising sowing time for newly-evolved short and medium duration genotypes under different agro-climatic regions.

(vii) Evaluating the response of clusterbean to micro-nutrients.

(viii) Studies on sources of S to achieve higher use-efficiency and B:C ratio.

(ix) Developing production technology of short-duration genotypes for maximising production in irrigated areas.

(x) Identification of critical stage (s) for life saving/ protective irrigation (s).

SUMMARY

Utilisation of land and higher total productivity could be enhanced through inclusion of clusterbean in intercropping system with cereals, oilseeds, pulses and other crops. The unbranched clusterbean genotype FS 277 was ideal for intercropping. First fortnight of July was the optimum sowing time for medium duration genotypes, whereas second fortnight of June was most appropriate for the short duration genotypes. A row spacing of 30 cm for unbranched and 45 cm for branched varieties was found suitable at most of the locations. However, an intra-row spacing of 15 cm for branched and 10 cm for unbranched genotypes, was optimum. Seed rate of clusterbean varied between 20-25 kg/ha depending upon varying agro-climatic conditions, genotypes, seed size and sowing time. The crop responded to 40 kg P_2O_5/ha at most of the locations, but at few locations the response was observed only upto 20 kg

P_2O_5/ha. The crop was also found to respond to 20 kg N/ha. The application of *Rhizobium* culture gave no response at majority of the sites. The application of 20 kg S/ha increased the seed yield, gum content and gum yield in sulphur deficient soils and proved economic. Soil application of 20-25 kg $ZnSO_4$/ha or spray of 0.5% $ZnSO_4$ also showed its effectiveness in Zn deficient soils. The most critical stage of crop-weed competition was between 15-30 days after sowing. Fluchloralin (basalin) was found effective for controlling weeds, for which 1.5 kg a.i./ha dose was most appropriate for medium and 0.75-1.0 kg a.i./ha for light soils. Maximum benefit of fertilizer application could be realized only when proper and timely weed-control measures were adopted.

REFERENCES

Ali, M., Dhar, S.N. and Hazra, C.R. 1978. Performance of clusterbean varieties on different dates of planting under dryland conditions of Bundelkhand. Abstract In : *'National Symp. on Non-monetary inputs in field crop prod.'* Feb. 11-13, 1978 held at HAU, Hisar.

Ali, M. 1982. Effect of plant types and row spacing on clusterbean production under dryland condition. *Indian J. Agron.* 27(2): 144-148.

Anonymous, 1976. Salient highlights of research conducted during 5 years period (1971-76). Coordinated Dryland Agriculture Research Project, HAU, Hisar, India.

Anonymous, 1996. *Ann. Progress Report - All India Coordinated Research Project on Arid Legumes.* (Coordinated unit Report).

Anonymous, 2000. *Ann. Progress Report* - Regional Research Station, Bawal, CCS HAU, Hisar, India.

Atwal, A.S. and Sidhu, G.S. 1964. Legumes in the nitrogen economy of soil. I Fixation and excretion of nitrogen by Indian legumes under controlled conditions in sand culture. *Indian J. agric. Sci.* **34** (3) : 139-45.

Bains, D.S and Dhillon, A.S. 1977. The influence of sowing dates and row spacing patterns on the performance of two varieties of clusterbean (*Cyamopsis tetragonoloba* L.). *J. Res. Punjab Agric. Univ.* **14** (2): 157-61.

Bhadoria, R.B.S., Chauhan, D.V.S. and Bhadoria, H.S. 1996. Effect of weed control on yield attributes and yield of clusterbean (*Cyamopsis tetragonoloba*). *Indian J. Agron.* **41** (4) : 662-4.

Bhadoria, R.B.S., Chauhan, G.S., Kushwaha, H.S. and Singh, V.N. 1992. Intercropping of clusterbean (*Cyamopsis tetragonoloba*) with pearl millet (*Pennisetum glaucum*). *Indian J. Agron.* **37**

(3) : 436-9.

Bhadoria, R.B.S., Tomer, R.A.S., Khan, H. and Sharma, M.K. 1997. Effect of phosphorus and sulphur on yield and quality of clusterbean (*Cyamopsis tetragonoloba*). *Indian J. Agron.* **42** (1) : 131-4.

Bhadoria, R.B.S. and Chauhan, D.V.S. 1994. Response of clusterbean (*Cyamopsis tetragonoloba*) to date of sowing and spacing. *Indian J. Agron.* **39** (1) : 156-7.

Bhardwaj, R.P., Bhargava, S.L. and Shankhla, H.C. 1982. Towards better harvest of *guar* in Rajasthan. *Indian Fmg.* **32** (6) : 10-2.

Bhati, D.S. and Manohar, S.S. 1989. Studies on the intercropping of pearl millet with legumes at two nitrogen levels. *Haryana J. Agron.* **5** (1) : 9-12.

Budhiraja, M., Rahtee, S.S., Yadav, A. and Panwar, R.S. 1998a. Effect of dinitroaniline herbicides on the yield and quality of clusterbean varieties. *Haryana J. Agron.* **14** (2) : 228-30.

Budhiraja, M., Rathee, S.S., Yadav, A. and Panwar, R.S. 1998. Sensitivity of clusterbean varieties to dinitroanilines in sandy loam soils.. *Haryana J. Agron.* **14** (2) : 127-9.

Chouhan, G.S. and Dungarwal, H.S. 1982. Production potential of maize (*Zea mays* L.) with different companion crops and their methods of planting. *Indian J. agric. Res.* **16** : 193-8.

Dahiya, S.K., Rana, V.S., Faroda, A.S. and Yadav, B.S. 1986. Response of clusterbean cultivars to nitrogen and phosphate fertilization under dryland conditions. *Haryana Agric. Univ. J. Res.* 16(2): 156-159.

Dahiya, S.S., Khera, A.P., Tomer, N.K., Grewal, K.S. and Pawar, B.S. 1986a. Response of fertilizers to *guar* under irrigation constraints. *Guar Res. Ann.* **4** : 18-20.

Das, B., Thareja, S.K. and Singh, K. 1977. Quality of guar seed as influenced by different agronomical practice. *Proceeding of the First ICAR Guar Research Workshop*, held during 11-12 January, 1977 at CAZRI, Jodhpur PP. 98-103.

Daulay, H.S, Henry, A. and Bhati, T.K. 1998. Compatibility of promising varieties of clusterbean for inter/mixed cropping in pearl millet in dryland of Western Rajasthan. *Forage Res.* **24** (1) : 13-6.

Daulay, H.S. 1977. Relative performance of annual grain legumes as companion crops in intercropping systems with perennial grasses. *Proc. of the First ICAR Guar Res. Workshop*, held during 11-12 January 1977 at CAZRI, Jodhpur, pp 93-7.

Daulay, H.S. and Singh, K.C. 1982. Chemical weed control in green gram and clusterbean. *Indian J. Agric. Sci.* **52** (11) : 758-63.

De, R., Rao, Y.Y. and Ali, W. 1983. Grain and fodder legumes as preceding crops affecting the yield and N economy of rice. *J. Agric. Sci.* **101** (2) : 463-6.

Deshpandey, P.M. and Nankar, J.T. 1977. Effect of intercropping on yield of seasonal sugarcane on net returns in Marathwara region. *Indian Sugar* **27** : 261-3.

Dhillon, G.S., Kler, D.S. and Walia, A.S. 1980. Studies on groundnut- clusterbean association. *J. Res. Punjab Agric. Univ.* **17** (1) : 34-7.

Dhukia, R.S, Singh, K.P. and Ram, S. 1990. Studies on the effect of sowing dates and irrigation levels on periodic accumulation of drymatter, leaf area index and seed yield of clusterbean (*Cyamopsis tetragonoloba* L. Taub.) varieties. *Guar Res. Ann.* **6** : 27-33.

Dhukia, R.S. and Singh, K.P. 1988. Effect of sowing dates and irrigation levels on grain quality of clusterbean. *Forage Res.* **14** (1) : 23-7.

Dixit, L. and Misra, A. 1991. Intercropping in Sugarcane. *Indian J. Agron.* 36 : 261-262.

Dutta, T.R, Gupta, J.N. and Gupta S.R. 1975. Pre and post emergence control of *Parthenium hysterophorus* L. *Sci. and Cult.* **42** : 179-81.

Dutta, T.R, Rao, C.N. and Verma, O.P.S. 1974. Preliminary evaluation of herbicides for forage cropping systems. *Ann. Report, IGFRI*, Jhansi, pp 96-7.

Gill, A.S., Patil, S.D. and Yadav, C.L. 1984. Intercropping studies in *Cenchrus ciliaris*. *Agril. Sci. Digest* 4 : 206-7.

Gill, P.S., Singh, K. and Paroda, R.S. 1974. Effect of irrigation and phosphorus levels on the yield of summer legumes. *Haryana Agric. Univ. J. Res.* **4** (3) : 186-9.

Gill, P.S. 1979. Response of clusterbean cultivars to phosphate, nitrogenous and bacterial fertilization in terms of growth, yield and gum production. *Thesis Abst.* **5**(3) : 163-9.

Gill, P.S. and Singh, K. 1981. Effect of fertilization on yield contributing characters and grain yield of clusterbean (*Cyamopsis tetragonoloba* L. Taub.) varieties. *Haryana Agric. Univ. J. Res.* **11** (3) : 333-8.

Gupta, V.K. 1983. Studies on the comparative response of some *kharif* crops to zinc application. *Agric. Sci. Dig.* **3** (2) : 79-80.

Hegde, D.M. and Randhawa, G.S. 1981. Economcis of nitrogen fertilization of intercropping system in medicinal yam (*Dioscorea floribunda* M). *Indian Drugs* **18** (9) : 322-6.

Jain, Veena, Joshi, U.N. and Taneja, K.D. 1988. Effect of phosphorus and Rhizobium culture on protein and gum content of clusterbean (*Cyamopsis tetragonoloba* L. Taub.). *Agric. Sci. Digest* **8** (1) : 9-11.

Jain, Veena, Yadav, B.D., Sharma, B.D. and Taneja, K.D. 1987. Effect of dates of sowing, row spacing and varieties on yield and quality of clusterbean (*Cyamopsis tetragonoloba* L. Taub.). *Indian J. Agron.* **32** (4) : 378-82.

Katyal, J.C. and Sharma B.D. 1979. Role of micronutrients in crop production. *Fertilizer News* **24** : 33-50.

Keshwa, G.L. and Singh H. 1991. Nitrogen management for production of pearlmillet (*Pennisetum glaucum*) and clusterbean (*Cyamopsis tetragonoloba*) forage mixture. *Indian J. Agron.* **36** (3) : 433-6.

Khan, H. and Guriar, B.S. 1997. Effect of phosphorus and sulphur on yield of clusterbean (*Cyamopsis tetragonoloba* L. Taub.). *Bharatiya Krishi Anusandhan Patrika* **12** (2) : 113-7.

Kumar, V., Yadav, A., Malik, R.K., Yadav, B.D. and Sachan, P.L. 1994. Performance and persistance of fluchloralin and pendimethalin applied in clusterbean. *Indian J. Weed Sci.* **26** (3&4) : 139-42.

Kumar, V., Yadav, B.D. and Agarwal, S.K. 1996. Performance of fluchloralin and pendimethalin to control weeds in clusterbean under rainfed condition. *Haryana J. Agron.* **12** (1) : 43-6.

Kumar, V., Yadav, H.D. and Dhankhar, O.P. 1999. Production Technology for Dryland in Southern Haryana. *A bulletin – CCS HAU Reg. Res. Sta.* Bawal (Haryana), India.

Kumar, V. and Agarwal, S.K. 1982. Studies on the compatibility and profitability of different intercrops in pigeonpea. *Haryana Agric. Univ. J. Res.* **12** (1) : 116-8.

Kumar, V. and Agarwal, S.K. 2000. Forage Production Technology in Southern Haryana. *A bulletin* – CCS Haryana Agricultural University, Hisar. India.

Maliwal, P.L., Singh, G.D. and Manohar, S.S. 1987. Effect of phosphorus and zinc levels on yield attributes and yield of clusterbean. *Rajasthan Agric. Univ. J. Res.* **1** (1) : 68-71.

Meena, K.C., Singh, G.D. and Mundra, S.L. 1991. Effect of phosphorus, micronutrients and irrigation on clusterbean. *Indian J. Agron.* **36** (2) : 272-4.

Menon, U. 1977. Genetic improvement of *guar* in Rajasthan. *Proc. First ICAR Guar Res. Workshop,* held during 11-12 January 1977 at CAZRI, Jodhpur, pp 48-52.

Misra, D.K. 1961. Role of legumes in crop husbandry of Western Rajasthan. *J. Soil Water Conser.,* **9** : 124-34.

Misra, D.K. 1964. Agronomic investigation in arid zone. Proc. of symposium on problems of Indian Arid Zone, Jodhpur, Ministry of Education, Govt. of India 165-9.

Mittal, S.P., Dabas, B.S., Thomas, T.A. and Chopra, D.P. 1977. Assessment of genetic stocks of guar (*Cyamopsis tetragonoloba* L. Taub.). Proc. First ICAR Guar Res. Workshop held during 11-12 January 1977 at CAZRI, Jodhpur, pp 58-68.

Muthuvel, P., Subramanian, V., Robinson, J.G. and Ravi, V. 1984. Soil fertility and crop yield under bajra-pulses intercropping system. *Madras Agri. J.* **71** (9) : 623-4.

Nandwal, A.S., Dabas, S., Bharti, S. and Yadav, B.D. 1990. Zinc effect on nitrogen fixation and clusterbean yield. *Ann. Arid Zone* **29** (2) : 99-103.

Narwal, S.S. and Malik, D.S. 1986. Effect of intercropping on growth and yields of rainfed sunflower and companion legumes. *Inter. J. Trop. Agri.* 4(1): 55-62.

Negi, L.S. and Singh, K. 1955. Effect of different rotations on yield of cotton. *Indian Cotton Growing Rev.* **9** (1) : 24-6.

Rai, R.K. and Sinha, M.N. 1993. Intercropping grain legumes involving pigeonpea, groundnut, urdbean, cowpea and *guar*. *Ann. Agri. Res.* **14** : 73-6.

Rajput, R.L., Bhadouria, S.S. and Tomer, S.P.S. 1989. Intercropping in pigeonpea. *Indian J. Agron.* 34(3): 373-5.

Rana, D.S., Yadav, B.D., Dhukia, R.S. and Midha, L.K. 1991. Effect of row spacing and intra-row spacing on seed yield of clusterbean under late sown conditions. *Guar Res. Ann.* **7** : 54-6.

Rana, V.S., Faroda, A.S., Dahiya, S.K. and Yadav, B.S. 1986. Effect of mixed/intercropping of mung bean and clusterbean in bajra (*Pennisetum typhoides*) under rainfed conditions. *Haryana Agri.Univ. J. Res.* **16** (1) : 71-6.

Rao, C.N., Sastry, J.A. and Verma, O.P.S. 1979. A note on the effect of cultural practices on yield of forage cereal-legume mixtures. *Forage Res.* **5** : 89-92.

Raut, M.S. and Ali, M. 1983. Response of clusterbean (*Cyamopsis tetragonoloba*) to phosphate application and Rhizobial inoculation under dryland conditions. *Legume Res.* **6** (2) : 65-8.

Reddy, G.S. and Venkateswarlu S. 1989. Effect of fertilizer and planting pattern in castor-clusterbean intercropping system. *J. Oilseeds Res.* **6** (2): 300-7.

Reddy, T.V., Reddy, K.R, Rajan, M.S. and Reddy, G.S. 1981. Production potential of seed clusterbean genotypes as a sequence crop in dryland. *Indian J. Agron.* **26** : 89-90.

Regmi, K.R., Sharma, R.N. and Pal, M. 1984. Yield and nutrient status of the soil as influenced by residual fertility left by clusterbean and direct fertilization of wheat in clusterbean-wheat system. *Indian J. Agron.* **29** (2) : 263-5.

Regmi, K.R., Sharma, R.N. and Pal, M. 1985. Dry matter accumulation and nutrient uptake in wheat as influenced by clusterbean-residual and direct fertilizaton of wheat. *Indian J. Agron.* 30(3): 254-6.

Rewari, R.B., Sen, A.N. and Sen, A. 1965. Nitrogen fixation by clusterbean (*Cymopsis tetragonoloba* L. Taub.) in relation to phosphate uptake from soil. *Indian J. Agri. Sci.* **35** (2) : 162-7.

Saxena, A., Singh, D.V. and Joshi, N.L. 1997. Effect of tillage and cropping systems on soil moisture balance and pearl millet yield. *J. Agron. and Crop Sci.* **178** : 251-7.

Sharma, B.D., Taneja, K.D., Kairon, M.S. and Jain, Veena 1984. Effect of dates of sowing and row spacing on yield and quality of clusterbean. (*Cyamopsis tetragonoloba* L. Taub.). *Indian J. Agron.* **29** (4) : 557-8.

Sharma, K.B., Singh, G.D. and Keshwa, G.L. 1988. Yield and return of pearlmillet and clusterbean intercropping as affected by nitrogen in semi-arid areas. *Haryana J. Agron.* **4** (2) : 135-7.

Sharma, P.K., Tiwana, M.S. and Gupta, B.K. 1993. Association of summer legume and non-legume fodders under rainfed conditions. *Agric. Sci. Digest* 13 : 20-2.

Shekhawat, P.S., Rathore, A.S. and Singh, M. 1996. Effect of source and level of sulphur on yield attributes and seed yield of clusterbean (*Cyamopsis tetragonoloba*) under rained conditions. *Indian J. Agron.* **41** (3) : 424-6.

Shivran, A.C., Khangarot, S.S., Shivran, P.L. and Gora, D.R. 1996. Response of clusterbean (*Cyamopsis tetragonoloba*) varieties to sulphur and phosphorus. *Indian J. Agron.* **41** (2) : 340-2.

Singh, Bharat, Singh, M. and Dang, Y.P. 1989b. Comparative performance of different zinc sources on maize and subsequent crop of guar (*Cyamopsis tetragonoloba*). *Haryana J. Agron* **5** (1) : 24-48.

Singh, B.P. 1981a. Seed yield, productivity, returns and water use efficiency as affected by planting system of guar and urdbean in dryland. *Guar Newsl.* **2**: 34-7.

Singh, B.P. 1981. Effect of intercropping of guar in pearlmillet. *Guar Newsl.* 2 : 24-7.

Singh, B.P. 1982. Seed yield and water use efficiency of guar variety HG 75 as affected by crop geometry in dryland. *Guar Newsl.* **3** : 9-11.

Singh, B.P. 1985. Effect of intercropping with pearlmillet on productivity, profitability and water use on aridisols. *Indian J. Agron.* **30** (4) : 408-13.

Singh, D P, Rathore D N, Singh H and Kumar V. 1978. A note on crude protein and gum production of two varieties of *guar* (*Cyamopsis tetragonoloba* L. Taub.) as influenced by different seed rates and row spacings. *Ann. Arid Zone* **17** (3) : 329-31.

Singh, G, Sindhu B S, Sahoo M S and Tiwana M S. 1980. *Fodder Crops. Cummuni. Centre*, PAU, Ludhiana. pp. 24-38.

Singh, G D, Keshwa G L and Sharma K B. 1989. Effect of nitrogen and intercropping of clusterbean on growth and yield of pearlmillet. *Haryana J. Agron.* **5** (2) : 179-81.

Singh, G. 1981 d. Effect of quality of irrigation water on the seed yield of *guar* under different soil textures. *Guar Newsl.* **2** : 34.

Singh, G. 1981c. Effect of intercropping on yield and economics of guar with and without basal dose of nitrogen and phosphorus. *Guar Newsl.* 2 : 38.

Singh, H. and Tiwana U.S. 1995. Response of guar (*Cyamopsis tetragonoloba* L. Taub.) varieties to varying levels of phosphorus and row spacing. *Indian J. agric. Res.* **29** (1) : 49-52.

Singh, J.P. and Singh B.P. 1988. Intercropping of green gram and *guar* in castor under dry land condition. *Indian J. Agron.* **33** : 177-80.

Singh, K., Kumar S. and Taneja K.D. 1979. Effect of different sowing dates on the seed yield of different varieties of guar (*Cyamopsis tetragonoloba* L.Taub.). *Haryana Agric. Univ. J. Res.* **9** (4): 312-6.

Singh, K.C. and Singh R.P. 1975. *Annual Progress Report* - Dry Farming Research Main Centre, CAZRI, Jodhpur India.

Singh, K.C. and Singh R.P. 1977. Intercropping of annual grain legumes with sunflower. *Indian J. agric. Sci.* **47** (11) : 563-7.

Singh, M. and Joshi N.L. 1994. Effect of mixed and intercropping systems on dry matter and grain yields of component crops in arid environment. *Ann. Arid Zone* **33** : 125-8.

Singh, M. and Joshi N.L. 1997. Pearlmillet based cropping system in relation to plant population and moisture availability. *Ann. Arid Zone* **36** (1) : 37-42.

Singh, M. and Joshi N.L. 1994. Performance of pearlmillet based intercropping system under drought conditions. Arid Soil Res. and Rehabilitation 8 : 313-19.

Singh, P.P. and Gautam D.P. 1973. Effect of double cropping on soil moisture and yield of barley under rainfed conditions. *Indian J. agri.Sci.* **43** : 988-92.

Singh, R.P. 1977a. Some agronomic considerations in the production of clusterbean. Proceedings of the First ICAR

Guar Res. Workshop held during 11-12 January at CAZRI, Jodhpur, pp. 75-92.

Singh, R.P. 1978. Guar Agronomy-A review article. In "*Guar*-its improvement and management". (R S Paroda and S K Arora Ed.,) *Forage Res.* **4A** : 41-8.

Singh, R.P. and Kaushik, M.K. 1997. Nitrogen economy in maize legume intercropping system. *Ann. Agri. Res.* **8** : 105-9.

Singh, R.P. and Tomer, R. 1975. *Ann. Progress Report* – Guar Agronomy. Dry Farming Research Main Centre, CAZRI, Jodhpur.

Singh, R.V. and Singh, R.R. 1989. Effect of nitrogen, phosphorus and seeding rates on growth, yield and quality of *guar* under rainfed conditions. *Indian J. Agron.* **34** (1) : 53-6.

Singh, R.V. and Singh, R.R. 1990. Uptake of nitrogen, phosphorus and potassium by guar (*Cyamopsis tetragonoloba* L. Taub.) as influenced by nitrogen with and without inoculation, phosphorus and seed rates under rainfed conditions. *Ann. Agri. Res.* **11**(3&4) : 329-32.

Singh, S.P. and Ahuja, K.N. 1990. Intercropping of grain sorghum with fodder legumes under dryland conditions in north western India. *Indian J. Agron.* **35** (3) : 287-96.

Singh, Ved, Sharma, S.K., Deo Ram, Siag R.K. and Verma, B.L. 1998. Performance of different crop sequences under various irrigation levels. *Indian J. Agron.* **43** (1) : 38-44.

Singh, Y.P., Dahiya, D.J., Kumar, V. and Singh, M. 1993. Effect of nitrogen application on yield and uptake of nitrogen by different legume crops. *Crop Res.* **6** (3) : 394-400.

Solanki, N.S., Shekhawat, R.P.S. and Shekhawat, B.S. 1998. Efficacy of phosphate culture with phosphorus levels on growth and yield of clusterbean (*Cyamopsis tetragonoloba* L. Taub.). *Ann. Arid Zone* **37** (4) : 417-8.

Sonani, V.V., Patil, R.R. and Patil, J.A. 1985. Critical period of crop weed competition in *guar* (*Cyamopsis tetragonoloba* L. Taub.). Paper abstracted In: *Annu. Conf. Indian Soc. Weed Sci.* pp 84.

Takkar, P.N. Mann, N.N. and Randhawa, N.S. 1973. Major rabi and kharif crops respond to zinc. *Indian Fmg.* 23(8) : 5-8.

Taneja, K.D., Bishnoi, O.P., Rao, V.U.M. and Ram, Niwas 1995. Effect of environment on growth and yield of clusterbean. *Crop Res.* **9** (1) : 159-62.

Taneja, K.D., Gill, P.S. and Sharma, B.D. 1982. Effect of row spacings and intra-row spacings on seed yield of *guar* cultivars. *Forage Res.* **8** (2) : 111-5.

Taneja, K.D., Saini, M.L. and Sharma, B.D. 1984. Effect of dates of sowing and row spacings on the seed yield of guar. *Forage Res.* **10** (2) : 115-7.

Teneja, K.D., Sharma, B.D. and Gill, P.S. 1981. Response of clusterbean to varying levels of phosphorus and bacterial culture. *Guar Newl.* 2 : 20-3.

Tiwana, U.S. and Tiwana, M.S. 1992. Effect of sowing time, seed rate and row spacing on the seed yield of guar (*Cyamopsis tetragonoloba* L.Taub) under rainfed conditions. *Forage Res.* **18** (2) : 151-3.

Tomer, Ramesh and Singh, R.P. 1976. *Ann. Report. Guar Agronomy*, Dry Farming Research Main Centre, CAZRI, Jodhpur.

Venkateswarlu, S. and Subramanian, V.B. 1990. Productivity of some rainfed crops in sole and intercrops system. *Indian J. Agric. Sci.* **60** : 106-9.

Yadav, B.D, Agarwal, S.K., Faroda, A.S. and Joon, R.K. 1990. Physiological basis of yield variation in clusterbean as affected by row spacing and fertilizer application. *Forage Res.* **16** (1) : 38-41.

Yadav, B.D., Agarwal, S.K. and Arora, S.K. 1989 a. Yield and quality of new cultivars of clusterbean as affected by row spacing and fertilizer application. *Guar Res. Ann.* **5** : 24-7.

Yadav, B.D., Agarwal, S.K. and Arora, S.K. 1989. Response of clusterbean to row spacing and fertilizer application. *Guar Res. Ann.* **5** : 8-13.

Yadav, B.D., Agarwal, S.K. and Faroda, A.S. 1991. Dry matter accumulation and nutrient uptake in clusterbean as affected by row spacing and fertilizer application. *Forage Res.* **17** (1) :39-44.

Yadav, B.D., Joon, R.K., Lodhi, G.P. and Sheoran R.S. 1991a. Effect of agro-management practices on the seed yield of clusterbean. *Guar Res. Ann.* **7** : 30-33.

Yadav, B.D., Joon, R.K. and Lodhi, G.P. 1997. Chemical weed control in clusterbean (*Cyamopsis tetragonoloba* L. Taub.). *Forage Res.* **23** (3&4) : 189-92.

Yadav, B.D., Joon, R.K. and Singh, V.P. 1997a. Crop-weed competition in clusterbean under rainfed conditions. *Haryana J. Agron.* **13** (2) : 92-6.

Yadav, B.D., Sharma, B.D. and Taneja, K.D. 1990a. Studies on weed control in clusterbean. Paper *Abst. in Intern. Symp. on Natural Reso. Manag. for a Sustainable Agri.* held during 6-10 February 1990, pp. 270.

Yadav, B.D. and Joon, R.K. 1993. Studies on inter cropping of pearl milet in *guar*. *Forage Res.* **19** (3&4) : 306-9.

Yadav, N.D. 1992. Studies on strip cropping of pearlmillet and arid legumes in rainfed arid western Rajasthan. *Curr. Agric.* 16 : 89-92.

Yadav, R.P., Sharma, R.K. and Shrivastava, U.K. 1997b. Fertility management in pigeon pea (*Cajanus cajan*)-based intercropping system under rainfed conditions. *Indian J. Agron.* **42** (1) : 46-9.

Yadav, R.S. 1998. Effect of weed removal in clusterbean under different rainfall situations in an arid region. *J. Agron. and Crop. Sci.* 181 (4) : 209-214.

Yadav, V.K., Yadav, B.D. and Joshi, U.N. 1993. Effect of weed control and fertilizer application on seed and gum yield of clusterbean under rainfed condition. *Forage Res.* **19** (3&4) : 341-2.

CHAPTER - 4

PHYSIOLOGY

B.K. Garg and Uday Burman

Guar is being grown in India, since ancient times for vegetable, cattle feed, green manure fodder etc. Of late, the crop has assumed great industrial importance because of the occurrence of gum in its endosperm, which constitutes 30-36 per cent of the whole seed. Guar gum (galactomannan gum), a highly mucilaginous material is used in textile, food processing, cosmetics, mining, explosive and oil industries. Despite its diverse uses, little research efforts have been taken up on plant physiological aspects influencing growth, yield and quality of clusterbean. Although clusterbean is known as an important drought resistant kharif legume but meagre information is available on its physiology and biochemistry, particularly under drought. Information available on physiological aspects, particularly under stress conditions, is outlined here.

Germination and Seedling Growth

Bolder seeds give higher germination compared to medium and small sized seeds. Increasing seed size have led to increased root and shoot length and higher dry matter production (Kalavathi and Ramamoorthy, 1992). Maximum germination normally occurs at 30°C in guar and it takes six days for complete germination (Zade *et al., 1990*). Storage of guar seeds for six months significantly reduced the germination percentage (Kalavathi and Ramamoorthy 1992). However, higher storage temperatures (16-35°C) coupled with high relative humidity may lead to complete loss of viability (Doijode 1989). Yadav and Pant (1979) also reported that detrimental effects of storage at high relative humidity are further aggravated at high temperatures. Takeshita and Konno (1985) found that vigour of the seeds stored for long time was negatively correlated with the glucose in exudates during 12 h soaking.

Guar seeds are known to germinate only alter sufficient imbibition, approximately twice their original weight. A constant temperature of 30°C, seed treatment with dilute sulphuric acid (Musil, 1946), scarification for five minutes (Hymowitz and Matlock, 1964) and sowing depth of 15 to 30 cm (Zheng *et al.,* 1980) are some of the practical means for obtaining prompt and uniform germination.

The endosperm of guar contains about 41% dry matter and 45% acetone insoluble solids (75% of this comprises galactomannan and 3-11% N and P). McClendon *et al.* (1976) in a five day germination study observed that all galactomannan and almost 95% dry matter was translocated from endosperm to cotyledons. In addition, N and P were also translocated. Ali and Karimullah (1958) also reported that galactose and mannose are converted into glucose and fructose during germination.

Presoaking of seeds at 25-30°C, reversed the low temperature mediated inhibition due to increase in the activity of certain isoenzymes of peroxidase (Zheng *et al.,* 1980). Presoaking in NaCl or solution of trace elements also increased germination of guar (Singh *et al.,* 1976). This has been attributed to mobilization of material into growing seedlings due to hydration-dehydration (Doijode and Raturi, 1987). Like many legumes, guar is also not able to germinate at very low soil water potentials. Likewise, high salinity and submergence significantly reduced seed germination in guar (Yadava *et al.,* 1975; Datta and Dayal, 1988).

Developmental Physiology

Exposure of guar seedlings to Hoagland solution alone or in combination with hot water treatment or sodium hypochlorite adversely affected primary root growth but increased the branching extensively (Stafford and McMichael, 1990). External environmental factors seem to regulate leaf development. Short day length and high temperature delayed the production of trifoliate leaf (Sparks and Posttethwait, 1967). Removal of colyledons of young seedlings promoted single leaf form. Application of 50 nanogram of morphactin also caused fusion of three leaflets of compound guar leaf to form simple leaf (Kathju and Tewari, 1973). In clusterbean, like most legumes, pod wall significantly contributes towards photosynthate requirement during seed filling stage. In this relation, it was reported that chloroembryo of guar fruit wall induced chloroplast pigment synthesis (Palanisamy, 1990).

Seed development in guar follows many complex changes. In the young lower bud, the ovule consists of nucellar tissue surrounded by two integuments, supported on a finicular stalk with the ovule. A single megaspore grows into the embryo sac. The megaspore undergoes mitotic division to form monosporic, 8-nucleate female geometophyte structure. After fertilization, the zygotes divides by a transverse wall to form basal and a terminal cell. The terminal cell develops into the embryo proper, while the basal cell forms the suspensor supporting the embryos. At the time of fertilization, nuclear type of endosperm is initiated by the fusion of a sperm nucleus and two polar nuclei. As the endosperm develops, the nucellus is broken down and nucellar tissue is no more present at seed maturity. The endosperm, which in most legumes is commonly digested and absorbed during embryo development, persists in the mature guar seed. Galactomannan gum extracted from guar seed is located in the endosperm. The outer integument is modified to form parts of seed coat, while the inner integuments in the mature seed is broken down or reabsorbed (Sabar *et al.*, 1956; Green *at al.*, 1964).

Mineral Nutrition

Guar does not require much nitrogen except a modest dose of 15-20 kg N ha^{-1}. The basal dressing of N imparts vigour to young seedlings and together with phosphorus provides a balanced nutrition to the crop. Atwal and Sidhu (1964) observed that small does of nitrogen in pot culture experiment significantly improved the nodulation, growth and nitrogen fixation of guar. Work conducted at several places have also indicated that 15-20 kg N ha^{-1} imparts vigour to the young seedling of guar. Fertilization of guar by 70 kg P or 25 kg N produced higher percentage of normal seedlings compared to unfertilized plant (Madalgeri and Rao, 1989). Foliar spray of N as nitrate increased seed germination (8-26%) and seedling vigour index (32-83%) of guar (Sathiyamoorthy and Vivekanandan, 1995) leading to higher yield.

Nitrate application influences nitrate redutase activity (NRA) in clusterbean. Nitrate upto 10 mM increased NRA but at higher concentrations of 150 and 200 mM, NR activity was depressed (Selvraj *et al.,* 1995). However, application of ammonia, amino acids and urea invariably reduced NR activity in guar. Nitrate reduction of clusterbean, on the other hand, could be in the roots of cluster bean (Nair *et al.*, 1988). It is pertinent to mention that clusterbean is amides producing legume as evident from the higher levels of amide in the xylem sap. Furthermore its nodules recorded lower activity of uricase and allatoinase (Sheoran *et al.,* 1982).

On soils, low in phosphorus, guar responds well to the application of phosphatic fertilizers. Application of 40 kg P_2O_5 ha^{-1} drilled with the seed alongwith a basal application of 20 kg N ha^{-1} provided economically high yields. Work carried out at this Institute has shown that guar responds to the application of phosphorus but the increase in seed yield obtained at levels beyond 40 kg P_2O_5 ha^{-1} was marginal.

Aggarwal and Praveen Kumar (1993) observed that phosphate application to legumes including clusterbean enhanced the per cent N in nodules and dry weight of nodules. Cultivation of legumes increased inorganic carbon by 12.5% over the initial level and with P application the level further increased from 21.2 to 28.7%. Kathju *et al.,* (1987) also found that P application led to an increase in soil N, particularly of the hydrolysable organic N fraction. Phosphorus application was also reported to increase elongated root systems besides, improving the growth and yield of guar (Tiwana and Tiwana, 1994).

Clusterbean responds to the application of zinc under dryland conditions with a optimum level of 5 kg ha^{-1}. In another study, under late rainfed sown situation at Durgapura, foliar spray of Zn (1 kg ha^{-1}) at 45 DAS increased the seed yield by about 30 per cent. Gandhi and Mehta (1959) found that boron deficient guar plant had thin stem, poor root development and reduced area of basal leaves. Similarly, copper deficient plants had young leaflets curled upwards along the margins followed by development of chlorotic areas (Neelkantan and Mehta, 1961). According to various studies the optimum levels of various essential nutrients have been presented in Table 1.

Table 1. Optimum level of essential nutrients for growth of clusterbean.

Nutrient element	Optimum level (ppm)
Nitrogen	400-450
Phosphorus	30-35
Potassium	400-450
Calcium	180-200
Magnesium	5-10
Sulphur	30-35
Boron	0.1-0.3
Copper	0.1-0.2

Source: Gandhi and Mehta, 1959; Neelkantan and Mehta, 1961

Biochemical Changes

Guar has higher phosphoenol pyruvate carboxylase (PEP carboxylase) activity and $^{14}CO_2$ fixation efficiency compared to mung bean (Kar *et al.,* 1991). Chloroembryos of guar have also been reported to have better dark fixation of CO_2 and O_2 evolution compared to leaf chloroplast (Kaladharan and Vivekanandan, 1990). Furthermore, its current photosynthesis was not the only pace regulator of symbiotic N fixation, because it was observed that decline in carbohydrate content in nodules was very slow and not commensurate with N_2-ase activity (Sheoran *et al.,* 1989). Variation in certain leaf metabolites due to growth/stage or varieties have also been reported in guar. Khatta *et al.* (1988) found varietal differences in amino acids profile, while Kaushal and Bhatia (1982) found variation in polyphenol content of leaves and seed depending on the stage of growth.

Plant Growth Regulators

Plant growth regulators are known to influence the plant metabolism of all crops including guar. However, their effect, invariably depends on the concentration and the stage of application. Seedling growth of guar was inhibited by cycocel but promoted by gibberlic acid. Simultaneous application of both the hormones led to reversal of inhibition caused by cycocel. GA promoted hypocotyl growth beyond control, whereas radical growth was not completely reversed. It also failed to reverse the inhibition of activities of P-glycerophosphatase acid phosphates and alkaline pyrophosphatase in germinating guar seedlings (Kathju *et al.,* 1971). Coupled with close spacing (10 x 6 cm) CCC gave higher guar yield (Rathore *et al.,* 1990). However, when coupled with 20 kg N ha^{-1} with 500 ppm CCC gave maximum green pod weight, number of seeds per pods, total protein percentage and low crude fibre percentage (Singh and Rajput, 1985) . B-9 at 2500 ppm also increased seed yield, 1000-seed weight and harvest index due to more accumulation of crude protein, soluble carbohydrates and starch but the gum content of seeds decreased progressively with increasing concentrations of B-9 (Uprety and Yadav, 1985). Foliar spray of growth retardant ethephon increased the growth of guar seedlings besides increasing the crude protein content (Verma and Sankhla, 1976). Paclobutrazol or XE-1019 on the contrary, retarded shoot elongation (Davis *et al*., 1987). Seed treatment gave different response depending upon its duration. Dianthane - M 45 (400-1200 ppm) for 12 h increased guar height and yield components compared to control. However, treatment for 24 h adversely affected the growth and yield of guar (Ramulu and Rao, 1992).

Drought Tolerance

Under dryland conditions, the growth and yield of arid legumes is often limited due to low precipitation and frequent droughts. Therefore, several studies have been conducted in the recent years to understand the physiology of drought tolerance in the tropical legumes. Studies have shown that increasing water stress delayed as well as reduced germination, decreased moisture uptake and adversely influenced the hypocotyl and radical growth of guar (Sheoran *et al.,* 1980). Water stress resulted in the accumulation of starch in seedlings that was more pronounced in the hypocotyl than in the radical. A highly significant negative correlation (r=-0.92) existed between the amount of starch and the growth of the embryoaxis. The activity of amylase and respiration rate decreased with water stress while there was no effect on total soluble carbohydrates.

Significant genotypic differences to water tress in clusterbean have been reported both at the germination (Sheoran *et al.,* 1980) and other growth stages as well (Kuhad and Sheoran, 1986, Stafford and McMichael, 1991, Garg *et al.,* 1998). These studies have indicated that clusterbean has a higher capacity to recover from water stress and thus, provides reasonable seed yield and dry matter once the stress is relieved.

Kuhad and Sheoran (1986) observed that water stress imposed at flower initiation stage in five clusterbean genotypes decreased relative water content (RWC), photosynthesis, starch content and carbohydrate accumulation. Furthermore, the activities of glutamine synthetase (GS), glutamate synthase (GOGAT) and glutamate dehydrogenase (GDH) were reduced in all the varieties. The seed yield and final dry matter production were reduced significantly in genotypes IC-11521 and FS-277. Garg *et al.* (1998) found that water stress at the flowering stage was most detrimental in genotypes Maru guar and FS-277 but pod developmental stages were most critical in other genotypes viz. Suvidha, HFG-182 and JGC-19 (Fig. 1). It was further observed that water stress induced decline in plant water potential and leaf RWC which led to decline in total chlorophyll and soluble protein besides accumulation of free proline to various extents depending the genotype and growth stage at which drought was experienced. Likewise, the activities of nitrate reductase, GS and GOGAT decreased but that of GDH increased indicating a possible shift in pathway of ammonia

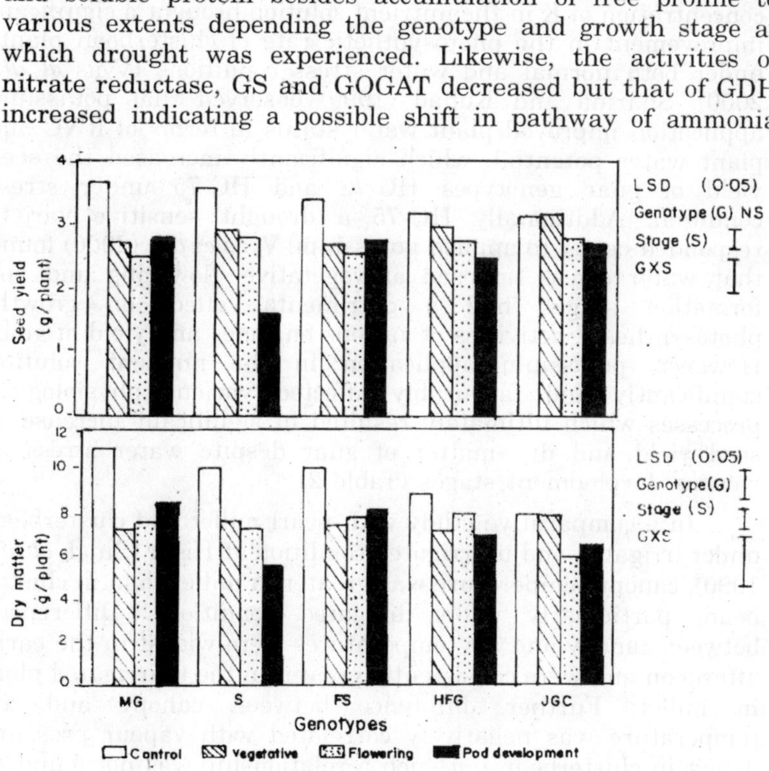

Fig. 1. Influence of drought at vegetative, flowering and pod development stages on seed and dry matter yield of clusterbean genotypes (MG-Maruguar; S-Suvidha, FS-277, HFG-HFG-182 and JGC-JGC-19) (Garg *et al.,* 1998).

assimilation from normal GS-GOGAT to GDH pathway under intense drought conditions as reported in moth bean crop (Vyas *et al.*, 1996).

Stafford and McMichael (1991) compared promising cultivars of clusterbean from USA, India and Pakistan under rainfed and irrigated conditions. The cultivars differed significantly in number of racemes per plant and number of pods per plant. However, the number of pods per plant, was the most affected yield component under water stress. In recent years, attempt have been made to alleviate the deterimental effects of drought on number of crops including clusterbean. In this regard application of potassium under soil moisture stress has been found to play a major role through its favourable influence on various physiological processes and yield (Mengal and Kirkby, 1980; Lahiri and Kackar 1985). Increasing concentration of K in the nutrient solution brought a significant improvement in the photosynthetic rate of clusterbean plants under both normal and water stress conditions (Vyas *et al.*, 2000). Sharma and Kuhad (1999) observed that potassium application improved plant water status in terms of RWC and plant water potential, which significantly increased the seed yield of guar genotypes HG-72 and HG-75 under stress condition. Additionally HG-75 a drought sensitive variety responded better to applied potassium. Vyas *et al.*, (2000) found that water stress imposed at vegetative, flowering and pod formation stages had a detrimental effect on growth, photosynthesis, activities of various enzymes and yield of guar. However, potassium application in the nutrient solution significantly and favourably affected various physiological processes which ultimately resulted in significant increase in seed yield and dry matter of guar despite water stress at various development stages (Table 2).

In a comparative study with pearl millet and clusterbean under irrigated and unirrigated condition at Hisar (Singh *et al.*, 1990), canopy temperature was greater in millet than in cluster bean, particularly under irrigated conditions. Differences between canopy and air temperatures were widest in the early afternoon and were consistently greater in the unirrigated plots in millet. Further difference between canopy and air temperature was negatively correlated with vapour pressure deficit in clusterbean but such a relationship was not found in pearl millet. Sekhon *et al.* (1996) have also observed that average canopy temperature was negatively correlated with biomass and leaf area index in both stressed and unstressed guar crops.

Table 2: Influence of potassium on seed yield and shoot dry matter of clusterbean grown under drought at vegetative (V), flowering (F) and pod formation (P) stages

K level	Seed yield (g plant^{-1})					Shoot dry matter (g plant^{-1})				
	C	V	F	P	Mean	C	V	F	P	Mean
Full	3.19	2.65	2.15	1.17	2.29	10.42	8.92	8.29	6.27	8.49
½	2.87	2.35	1.98	1.26	2.11	10.10	9.08	8.03	6.02	8.31
¼	2.84	2.41	1.86	1.15	2.07	9.23	8.37	7.83	6.52	7.99
1/8	2.59	2.12	1.84	1.00	1.89	9.09	8.19	7.73	6.11	7.78
Mean	2.87	2.38	1.96	1.15	–	9.71	8.65	7.97	6.23	–
LSD		0.05		0.01			0.05		0.01	
Drought (D)		0.30		0.39			0.69		0.90	
Potassium (K)		0.30		0.39			0.69		0.90	
DxK		0.60		0.79			1.37		1.79	

Source: (Vyas *et al.,* 2001) C- Control.

It has been observed that early flowering/maturing genotypes of clusterbean perform better than late flowering/ maturing genotypes under low rainfall conditions. However, late genotypes out yielded early types under high rainfall conditions. The studies with early and late maturing genotypes adopting different planting patterns revealed that planting of early (RGC-936) and late (Maru guar) genotype in 1:1 ratio (either as alternate rows or as a mixture by weight) provided high and stable yields under both low and high rainfall conditions (Vyas *et al.,* 1997). This was attributed to better utilization of the limited available moisture and higher photosynthetic efficiency under genocropping as compared to sole cropping system.

Salt Tolerance

In view of the potential of cultivation of clusterbean on saline drylands (Lahiri *et al.,* 1987), a number of studies have been conducted to understand the physiology of salt tolerance of this crop and the subject has been reviewed recently (Garg and Gupta, 1997).

Several studies indicate the depressive effects of salinity on seed germination and early seedling growth of clusterbean (Mehta and Desai, 1958, Maliwal and Paliwal, 1982; Datta and Dayal, 1988). However, in most of these studies clusterbean was found tolerant to salt stress at germination stage. Salt levels greater than 8.5 dSm^{-1} delayed but did not significantly reduce the final emergence percentage upto 18.8 dSm^{-1}. Significant varietal differences to soil salinity at germination stage reported by Yadava *et al.* (1975) and Datta and Dayal (1988) suggested that there existed sufficient variability to salt stress and tolerant genotypes may be selected at the seedling stage.

Most of the available reports indicate that clusterbean is moderately salt tolerant crop and a significant reduction in growth and seed yield occurs only after a threshold level of salinity is reached (Mehta and Desai, 1958; Varshney, 1980. Lahiri *et al.* (1987) observed that final growth and yield of guar significantly decreased at 6 dSm^{-1}. Furthermore, reduction in seed yield was not associated with a significant decline in seed weight, indicating that salinity influenced yield primarily through growth reduction while the photosynthate transport *per se* remained relatively unaffected (Lahiri *et al.,* 1987). Irrigation with saline waters upto 80 mel^{-1} salt concentration, at the flowering stage did not also significantly affect final plant height, dry matter and seed yield of clusterbean (Garg *et al.,* 1986). However, at 120 me 1^{-1} all these parameters were significantly decreased. Salinity has also been reported to reduce the galactomannan content of guar seed endosperm (Kumar and Sharma, 1989).

Table 3. Effect of soil salinity (10 dSm-1) on seed yield, dry matter production and leaf area (at 40 DAS) of tolerant and sensitive clusterbean genotypes

Genoptype	Seed yield ($g\ plant^{-1}$)			Dry matter ($g\ plant^{-1}$)			Leaf area ($cm^2\ plant^{-1}$)		
	C*	S**	% reduction	C*	S**	% reduction	C*	S**	% reduction
Tolerant									
Malosen	3.83	3.27	14.6	10.0	08.0	10.8	21.3	190	11.1
HFG-182	3.27	2.73	16.5	09.0	08.0	10.7	228	195	14.5
Sensitive									
Durga Jai	3.88	1.92	50.5	08.8	05.4	38.7	26.1	120	62.0
FS-277	5.63	3.02	46.4	10.7	07.8	26.5	358	205	43.0
LSD (0.05)									
Genotype		0.69			1.7			27	
Salinity		0.32			0.8			12	
GxS		0.95			2.6			38	

Source: Lahiri *et al.,* 1996 *C-Control, **S-Saline

The tolerant genotypes (Malosen and HFG-182) did not experience more than 20% reduction in seed yield and dry matter at 10 dSm^{-1} level of soil salinity (Table 3). On the contrary, sensitive genotypes (Durga Jai and FS-277) suffered 40-50% reduction in growth and yield at the equivalent salinity (Lahiri *et al.,* 1996). Varshney (1980), also found that plant height and leaf number were reduced with increasing salinity (0.8 and 16 dSm^{-1}) in cv. 5196 but were stimulated in cv. S2395 at 8 dSm^{-1} and remained unaffected even at 16 dSm^{-1}. Garg *et al.* (1997) found that sensitivity to salt varied with the growth stages and the flowering stage was the most sensitive to NaCl salinity. However, dry matter production was more adversely affected due to salinity at the seedling stage, followed by vegetative and flowering stages. Garg *et al.*, (1986) observed that increasing salinity of irrigation water caused a progressive decline in plant water potential and RWC and an increase in leaf diffusive resistance of clusterbean plants. However, the differences were significant only at higher level of salinity. Varshney (1980) also observed that increasing salinity decreased tissue moisture content in two clusterbean genotypes. Garg *et al.*, (1996) found that $NaHCO_3$ salt was most detrimental to shoot dry weight, pod and seed yields, followed by Na_2SO_4 and NaCl. At 10 dSm^{-1} salinity, the reduction in seed yield was 53.2, 48.7 and 45.0 per cent under $NaHCO_3$, Na_2SO_4 and NaCl salinity, respectively, compared to control. Likewise, the corresponding decrease in shoot dry weight was 47.5, 39.6 and 33.7 per cent. Relatively few workers have investigated the mineral composition of guar under salt stress. Mehta and Desai (1958) observed that concentrations of calcium, sodium, and chloride increased and those of potassium and magnesium decreased with increasing NaCl concentrations. Lahiri *et al.,* (1987) observed that uptake of N, P and K declined with increasing soil salinity at different developmental stages although ageing in all the cases increased the uptake. Similarly, Garg *et al.,* (1997) observed that increasing NaCl concentration (0, 50, 100 and 150 mm) progressively decreased the concentrations of K and Ca and increased the concentration of Na in the shoots. Supplemental calcium (2.5 and 5.0 mM) enhanced Ca and K contents and reduced the concentration of Na thereby increasing the K: Na ratio at all levels of NaCl. Lahiri *et al.* (1996) found significant genotypic differences in mineral composition of salt tolerant and sensitive genotypes of clusterbean. Sensitive genotypes Durga Jai and FS-277 had higher concentrations of Na and Cl in the shoot tissue and a lower K : Na ratio under salinity stress as compared to tolerant genotypes Malosen and HFG-182.

The inhibition of growth and development of plants under salt stress is known to be often associated with many metabolic disturbances. In clusterbean also salt induced metabolic impediments have been reported by several investigators. Lahiri *et al.* (1987) observed that the adverse effects of soil salinity on the levels of total chlorophyll, starch, reducing sugars, soluble protein, free amino acids and free proline were discernible at 4 dSm^{-1} onwards. The derangement in protein metabolism was further obvious from the enhanced activity of protease and reduced activity of nitrate reductase with increasing soil salinity. The activities of peroxidase as well as those of acid and alkaline pyrophosphatases also increased progressively with increasing salinity. Garg *et al.* (1986) also reported similar responses on the levels of starch, reducing sugars and free proline and activity of nitrate reductase in clusterbean plants irrigated with increasing concentration of saline waters. Garg *et al.* (1996) further observed that different sodium salts had variable effects on leaf metabolism of clusterbean. Na_2SO_4 increased the chlorophyll content but $NaHCO_3$ and NaCl decreased it at similar salinity levels (5 and 10 dSm^{-1}). However, RNA and free proline were increased by all the salts and maximum RNA was under Na_2SO_4. This indicates that leaf metabolites and enzyme activities are affected differently depending on the type of salinity as also reported in several other studies.

Lahiri *et al.* (1996) found that sensitive genotypes consistently displayed higher metabolic alterations than tolerant genotypes in all the biochemical parameters studied. For instance, starch content did not decline in tolerant genotypes Malosen and HFG-182 but decreased significantly in sensitive varieties Durga Jai and FS-277. Furthermore, in sensitive genotypes showing more than 40% decline in starch concentration, reducing sugars also increased by 41 to 43%. The activities of NR and GS were significantly reduced in Durga Jai and FS-277 but were hardly affected in Malosen and HFG-182. Garg *et al.*, (1997) reported detrimental effects of increasing NaCl concentration on soluble protein, free amino acids and enzymes of nitrogen metabolism (viz., NR, GS, GOGAT and GDH). However, application of $CaCl_2$ significantly alleviated these adverse effects of salinity (Fig. 2). The results indicated that elevated calcium levels could protect the tissues from NaCl toxicity at the sub-cellular level besides improving nutrient uptake under saline conditions.

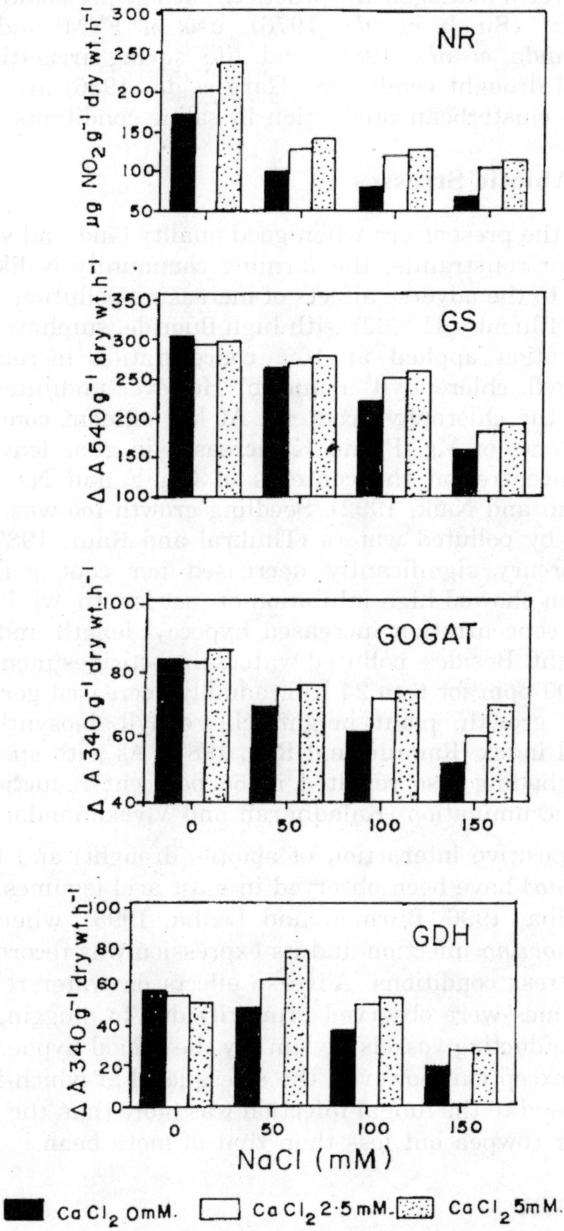

Fig. 2. Influence of supplemental calcium on the activities of NR, GS, GOGAT and GDH of clusterbean grown under different levels of NaCl salinity (Garg *et al.*, 1997).

Several management practices such as pre-soaking of seed treatment (Singh *et al.,* 1976), use of FYM and gypsum (Manchanda *et al.,* 1988) and life saving irrigation under terminal drought conditions (Garg *et al.*, 1986) are known to enhance clusterbean production in saline conditions.

Other Abiotic Stresses

In the present era when good quality land and waters are the major constraints, the farming community is likely to be exposed to the adverse affects of increasing pollution. Fertilizer factory effluent (pH 2.62) with high fluoride, sulphate and solid concentration applied at 10% concentration increased total chlorophyll, chlorophyll 'a' and 'b'. However undiluted effluent lowered the chlorophyll content. At low effluent concentration the contents of Mg, P and K increased in guar leaves but at high concentration the contents of Ca, F and Na increased (Goswami and Naik, 1992). Seedling growth too was adversely affected by polluted waters (Thukral and Kaur, 1987), Cobalt and mercury significantly decreased per cent germination. Cadmium showed high inhibition of root length, while only Mn at high concentration increased hypocoyl length and seedling dry weight. Besides, polluted waters, insecticides monocrotophs (300-1200 ppm for 6 to 24 h) gradually decreased germination, seedling growth, plant height, chlorophyll, biosynthesis and stomatal index (Ramulu and Rao, 1987). As with specific ions, simple shading also resulted in 65 per cent reduction in dry matter accumulation (Kaladharan and Vivekanandan, 1990).

A positive interaction of abiotic (drought) and biotic (*M. phaseolina*) have been observed in most arid legumes (Burman and Lodha, 1996, Burman and Lodha, 1998), whereas more *Macrophomina* infection and its expression was recorded under water stress conditions. Adverse effects on water relations of the legumes were observed primarily due to plugging of main water conducting vessels (xylem) by the fungal hypae. Guar too was no exception, however, the stress level at which it became predisposed to the fungal infection was more than the threshold value for cowpea but less than that of moth bean.

SUMMARY

Clusterbean (*Cyamopsis tetragonoloba*) is one of the major drought hardy *kharif* crops of dry lands and is grown for vegetable, cattle feed, green manure, fodder and various industrial uses because of the occurrence of gum in its endosperm. The available information on physiological aspects of seed germination, growth and development, mineral nutrition, drought and salt tolerance of this crop has been reviewed in view of its wide cultivation in arid and semi-arid regions. Higher germination of clusterbean takes place at an optimum temperature of 30°C after sufficient imbibitions and by using larger seeds. Seed treatment with dilute H_2SO_4, scarification for five minutes and sowing depth of 15-30 cm provides uniform germination. Being a leguminous crop, guar requires little nitrogen application but it responds significantly to phosphorus application up to 40 kg ha^{-1} in soils low in P supply. Zn application is also known to enhance growth and yield of guar in dry lands. Although clusterbean is fairly drought tolerant crop but significant genotypic differences to water stress observed at the germination and other growth stages indicate the prospects of further improvement of its drought tolerance capacity. Among many yield components affected by water stress, number of pods per plant appears most important trait contributing forwards yield. Maintenance of adequate potassium in soil improves plant water relations, photosynthesis and yield of clusterbean. Early genotypes of clusterbean perform better under low rainfall conditions while late genotypes out yield the early types under good rainfall conditions. However, planting of early and late genotypes in 1:1 ratio in alternate rows has been found most appropriate for obtaining sustainable yields under diverse rainfall situations. Seed treatment with thiourea followed by foliar spray at 25 and 40 days after sowing significantly alleviate the detrimental effects of field drought in clusterbean crop.

Guar is also moderately tolerant to salinity and significant reduction in yield occurs only after a threshold level of salinity (6 dSm^{-1}) is reached. Flowering stage has been found to be most sensitive to salinity. Salt stress depresses the concentrations of N, Mg, K and P whereas significantly increases those of Ca, Na and Cl. Supplemental calcium alleviated the detrimental effects of NaCl salinity by decreasing Na absorption, enhancing Ca and K uptake and favourably influencing nitrogen metabolism. Tolerant genotypes of clusterbean experience less metabolic alterations than sensitive genotypes due to differential absorption of Na and K. Several management practices have been discussed for yield improvement of clusterbean under abiotic stresses.

REFERENCES

Aggarwal, R.K. and Praveen Kumar, 1993. Changes in soil nitrogen due to cultivation of green legumes fertilized with phosphate in an arid loamy sand soil. *Ann. Arid Zone* **32** : 195-196.

Ali, G. and Karimullah, 1958. Germination of seeds of four (*Cyamopsis tetragonoloba*): Chemistry and analysis. *Pakistan J. Sci. Res.* **10** : 93-97.

Atwal, A.S. and Sidhu, G.S., 1964. Legumes in nitrogen economy of soil.1. Fixation and excretion of nitrogen by Indian legumes under controlled conditions in sand culture. *Indian J. agric. Sci.* **34** : 139-145.

Burman, U. and Lodha, S. 1996. *Macrophomina phaseolina* induced changes in plant water relations of resistant and susceptible cowpea genotypes. *Indian Phytopath.* **49** : 254-259.

Burman, U. and Lodha, S. 1998. *Macrophomina phaseolina* infection and water relation in mungbean genotypes. *Indian Phytopath.* **51** : 367-369.

Datta, K.S. and Dayal, J., 1988. Effect of salinity on germination and early seedling growth of guar. (*Cyamopsis tetragonoloba*). *Indian J. Pl. Physiol.* **31** : 357-363.

Davis, T.D., Gehlot, H.S., Williams, C.F. and Sankhla,, N., 1987. Comparative shoot growth retarding activities of paclobutrazol and Xe 1019. *Proc. Pl. Growth Reg. Sco. A:.* 121-124.

Doijode, S.D., 1989. Deteriorative changes in clusterbean seeds stored in different conditions. *Veg. Sci.* **16** : 89-92.

Doijode, S.D. and Raturi, G.B., 1987. Effect of hydration and dehydration on the germination and vigour of certain vegetable seeds. *Seed Res.* **15** : 156-159.

Garg, B.K., Kathju, S., Vyas, S.P. and Lahiri, A.N., 1986. Effect of saline waters on drought affected clusterbean. *Proc. Indian Acad. Sci. (Pl. Sci.)* **96** : 531-538.

Garg, B.K., Kathju, S., Vyas, S.P and Lahiri, A.N., 1997. Sensitivity of clusterbean to salt stress at various growth stages. *Indian J. Pl. Physiol.* **2** : 49-53.

Garg, B.K., Kathju, S., Vyas, S.P. and Lahiri, A.N. 1997. Alleviation of sodium chloride induced inhibition of growth and nitrogen metabolism of clusterbean by calcium. *Biol. Plant.* 39 : 395-401.

Garg, B.K., Kathju, S., Vyas, S.P. and Lahiri, A.N., 1996. Relative effects of sodium salts on growth, yield and metabolism of arid legumes. *Pl. Physiol. Biochem.* **23** : 148-152.

Garg, B.K., Kathju, S., Vyas, S.P. and Lahiri, A.N., 1998. Influence of water deficit stress at various growth stages on some enzymes of nitrogen metabolism and yield in clusterbean genotypes. *Indian. J. Pl. Physiol.* **3** : 214-218.

Garg, B.K. and Gupta, I.C., 1997. Physiology of salt tolerance of Arid Zone Crops II. Clusterbean. *Curr. Agric.* **21** : 23-24.

Gandhi, S.C. and Mehta, B.V., 1959. Studies on boron deficiency and toxicity symptoms in some common crops of Gujarat. *Indian J. agric. Sci.* **29** : 63-70.

Goswami, M. and Naik, M.L., 1992. Effect of fertilizer effluent on chlorophyll contents of *Cyamopsis tetragonoloba* (L) Taub. *J. Environ. Biol.* **13** : 169-174.

Green, M.J., Sparks, P.N. and Postlethwait, S.N., 1964. Studies of the ovule and seed development of guar. *Proc. Indian Acad. Sci.* **73** : 97-104.

Hymowitz, T. and Matlock, R.S., 1964. Guar: Seed, Plant and Population Studies. *Okla Agric. Exp. Sta. Tech. Bull, (B)* **108** : 1-35.

Kaladharan, P. and Vivekanandan, M., 1990. Phytosynthetic potential and accumulation of assimilate in the developing chloroembryos of *Cyamopsis tetragonoloba* (L) Taub. *Pl. Physiol.* **92** : 408-412.

Kalavathi, D. and Ramamoorthy, K., 1992. A note on the effect of seed size on viability and vigour of seed in clusterbean (*Cyamopsis tetragonoloba* (L) cultivar Pusa Navbhahar. *Madras Agric. J.* **79** : 530-532.

Kaushal, G.P. and Bhatia, I.S., 1982. A study of polyphenols in the seeds and leaves of guar. (*Cyamopsis tetragonoloba*) *J. Sci. Food & Agric.* **33** : 461-470.

Kar, M., Mythili, J.B. and Nair, T.V.R., 1991. Carbondioxide fixation in the nodules of amide transporting vs ureide transporting legumes *J. Nuclear Agric. and Biol.* **19** : 257-260.

Kathju, S. and Tewari, M.N., 1973. Morphogenetic responses of clusterbean plants to growth regulators. *Acta. Bot. Acad. Sci. Hungary* **19** : 189-193.

Kathju, S., Aggarwal, R.K. and Lahiri, A.N., 1987. Evaluation of diverse effects of phosphate application on legumes of arid areas. *Trop. Agric.* **64** : 91-96.

Kathju, S., Tewari, M.N. and Chatterji, U.N., 1971. Study of the effect of gibberlic acid and cycocel on the activity of phosphatases in guar *Z. Pflanzenphysiol.* **4** : 163-274.

Khatta, V.K., Kumar, N. and Gupta, P.C., 1988. Chemical composition and aminoacid profile of four varieties of guar (*Cyamopsis tetragonoloba*) seed. *Indian J. Animal Nutr.* **5** : 325-326.

Kuhad, M.S. and Sheoran, I.S., 1986. Physiological and biochemical changes in clusterbean genotypes under water stress. *Indian J. Pl. Physiol.* **29** : 46-52.

Kumar, A. and Sharma, B.K., 1989. Note on salinity induced changes in galactomannan gum content in endosperm of guar seeds. *Adv. Pl. Sci.* **2** : 300-301.

Lahiri, A.N. and Kackar, N.L.,1985. Influence of potassium on plants under soil water deficit. PR. II *Res. Rev. Series* **2** : 57-66.

Lahiri, A.N., Garg, B.K., Kathju, S., Vyas, S.P. and Mali, P.C., 1987. Responses of clusterbean to soil salinity. *Ann. Arid Zone* **26** : 33-42.

Lahiri, A.N., Garg, B.K., Kathju, S., Vyas, S.P. and Mali, P.C., 1996. Genotypic differences to soil salinity in clusterbean. *Arid Soil Res. and Rehab.* **10** : 333-345.

Madalgeri, M.S. and Rao, M.M., 1989. Effect of fertilizer levels and staggered removal by early produced fresh pod on the quality of Pusa Navbahar clusterbean (*Cyamopsis tetragonoloba*) seeds. *Seeds and Farms* **15** : 7-10.

Maliwal, G.L. and Paliwal, K.V., 1982. Salt tolerance of some mung bean (*Vigna radiata*), urad bean (*V. mungo*) and guar (*Cyamopsis tetragonoloba*) varieties at germination and early growth stages. *Legume Res.* **5** : 23-30.

Manchanda, H.R., Sharma, S.K., Malik, H.R. and Suneja, B.K., 1988. Technology for using sodic water for guar and bajra. *Indian Fmg.* **38** : 11-15.

McClendon, J.H., Nolan, W. and Wenzier, H.F., 1976. The role of endosperm in germination of legumes. Galactomannan, nitrogen and phosphorus changes in the germination of guar. *American. J. Bot.* **3** : 790-797.

Mehta, B.V. and Desai, R.S., 1958. Effect of soil salinity on germination of some seeds. *J. Soil Water Conserv. India* **6** : 169-176.

Mengel, K. and Kirkby, E.A., 1980. Potassium in crop production. *Adv. Agron.* **33** : 59-110.

Musil, A.F. 1946. The germination of guar (*Cyamoposis tetragonoloba* (L) Taub. *J. American Soc. Agron.* **38** : 661-662.

Nair, T.V.R., Karim, M.S., Pande, H.C. and Mythili, J.B. 1988. Clusterbean accumulates nitrate in roots. In : Proc. *Int. Congr. Pl. Physiol.* Vol. 2 pp. 1065-1070., (Eds. Sinha, S.K., Sane, P.V., Bhargava, S.C. and Agarwal, P.K.), New Delhi.

Neelkantan, V. and Mehta, B.V. 1961. Studies on copper deficiency and toxicity symptoms in some common crops of Gujarat (India). *J. Agric. Sci.* **56** : 293-298.

Palanisamy, K. 990. Can endosperm and fruit wall induce chloroplast pigment synthesis in chloroembryos of angiosperms. *Photosynthetica* **24** : 261-265.

Ramulu, C.A. and Rao, D. 1987. Effect of monocrotophos on seed germination, growth and chlorophyll content of clusterbean. *Comparative Physiol. Ecol.* **12** : 102-105.

Ramulu, C.A. and Rao, D. 1992. Influence of Dianthane M-45 on growth, flowering, fruiting and seed yield in guar (*Cyamoposis tetragonoloba* (L) Tabu). *Adv. Pl. Sci.* **5** : 12-16.

Rathore, S.V.S., Puneet Kumar, Singh, D.R. 1990. Effect of intra-row spacing and cycocel on the performance of seed crop of vegetable clusterbean: a note. *Haryana J. Hort. Sci.* **19** : 219-221.

Saber, A.H., Ahmed, Z.F. and Dararih, M. 1956. A contribution to the study of guar seeds grown in Egypt. *Bull Inst. Desert Egypt.* **6** : 67-78.

Sathiyamoorthy, L. and Vivekanandan, M. 1995. Vigour of *Cyamopsis tetragonoloba* (guar) seed produced from foliar spray of nitrogenous salts under rainfed conditions. *Legume Res.* **18** : 50-52.

Sekhon, N.K., Thind, S.S., Sidhu, R.S. and Aggarwal, G.C. 1996. Water status of green manure crop as affected by moisture stress at different growth stages. *Ann. Agric. Res.* **17** : 152-157.

Selvraj, K., Hemlatha, C. and Ghandhi, S.J. 1995. Differential nitrate reductase activity in NO_3 grown clusterbean, green gram-inhibiting effects of ammonia, aminoacids and urea. *Indian J. Expt. Biol.* **33** : 456-458.

Sharma, K.D. and Kuhad, M.S. 1999: Influence of potassium on plant water relations and yield of clusterbean under soil moisture stress. In: *Management of Arid Ecosystem*. (Eds. A.S. faroda, N.L. Joshi, S. Kathju and Amal Kar) pp. 267-270. Arid Zone Res. Assoc. of India and Scientific Publishers, Jodhpur.

Sheoran, I.S., Khan, M.I. and Garg, O.P. 1980. Differential behaviour of guar genotypes to water stress during germination. *Guar Newsl.* **1** : 3-4.

Sheoran, I.S., Luthra, Y.P., Kuhad, M.S. and Singh, R. 1982. Clusterbean a ureide or amide producing legume. *Pl. Physiol.* **70** : 917-928.

Sheoran, I.S., Swaraj, K., Dabas, S. and Hooda, R.S. 1989. Acetylene reduction activity as affected by detopping in clusterbean (*Cyamopsis tetragonoloba*) (L). *Indian J. Expt. Biol.* **27** : 746-748.

Singh, S.J.P. and Rajput, C.B.S. 1985. Effect of nitrogen,

phosphorus and cycocel on physicochemical characters of clusterbean vegetables cv. Pusa Navabahar. *Progressive Hort.* **17** : 181-184.

Singh, M., Lal, P. and Singh, K.S. 1976. A preliminary study on the indudcement of salt tolerance in guar (*Cyamopsis tetragonoloba*) var. Chikna at germination stage. *Agrochemica* **20** : 88-92.

Singh, D., Rao, V.V. and Bishnoi, O.P. 1990. Canopy temperature in pearl millet and clusterbean under irrigated and unirrigated conditions. *Haryana Agric. Univ. J. Res.* **20** : 23-27.

Sparks, P.N. and Posttlethwait, S.N. 1967. Comparative morphogenesis of dimorphic leaves of *Cyamopsis tetragonoloba*. *Am. J. Bot.* **54** : 281-285.

Stafford, R.E. McMichael, B.L. 1990. Primary root and lateral root development in guar seedling: *Environ. Exp. Bot.* **1** : 27-34.

Staffrord, R.E. McMichael, B.L. 1991. Effect of water stress on yield components in guar. *J. Agron. & Crop Sci.* **166** : 63-68.

Takeshita, H. and Konno, S. 1985. Application of Tes. Tape method to examine viability of stored legume seeds. JARQ **19** : 151-152.

Thukral, A.K. and Kaur, P. (1987). Effect of some trace elements of polluted waters on the germination of *Cyamopsis tetragonoloba* Taub. *Indian J. Ecology* **14** : 185-188.

Tiwana, U.S. and Tiwana, M.S. 1994. Effect of source, level and method of phosphorus application on the seed yield of clusterbean. *Indian J. Ecol.* **21** : 117-121.

Uprety, M. and Yadav, R.B.L. 1985. Influence of growth retardant (B-9) on growth, flowering, fruiting and seed yield of guar plants under pot culture. *Indian J. Pl. Physiol.* **28** : 1282-1289.

Varshney, K.A. 1980. Growth and development of two differentially salinized guar varieties under the influence of some hormones. *Indian J. Pl. Physiol.* **23** : 199-205.

Verma, C.H. and Sankhla, N. 1976. Effect of ethephon on growth and metabolism of *Cyamopsis tetragonoloba* (L.) Taub. *Forage Res.* **2** : 41-48.

Vyas. S.P., Garg, B.K., Kathju, S. and Lahiri, A.N. 1997. Physiological studies on genotypic management of clusterbean for sustainable production. In: *Recent Advances in Management of Arid Ecosystem*. (Eds. Kathju, S. Abichandani, R.K., Kar, A. and Joshi, N.L.) p. 53. Arid Zone Res. Asso. of India, Jodhpur.

Vyas, S.P., Garg, B.K., Kathju, S. and Lahiri, A.N. 2001. Influence of potassium on water relation, photosynthesis, nitrogen metabolism and yield of clusterbean under soil moisture deficit stress. *Indian J. Pl. Physiol.* **6** : 30-37.

Vyas, S.P., Kathju, S., Garg, B.K. and Lahiri, A.N. 1996. Response of clusterbean genotypes to shade. *Indian J. Pl. Physiol.* **1** : 234-238.

Vyas, S.P., Kathju, S., Garg, B.K. and Lahiri, A.N. 1996. Activity of nitrate reductase and ammonia assimilating enzymes of moth bean under water stress. *Sci. & Cult.* **62** : 213-214.

Yadav, T.D. and Pant, N.C. 1979. Moisture content-relative humidity relationship of legume seeds. *Seed Res.* **7** : 11-17.

Yadava, R.B.R., Mehra, K.L., Magoor, M.L., Sreenath, P.C. and Yadav, M.S. 1975. Varietal differences in salt tolerance during seeds germination of guar. *Indian J. Pl. Physiol.* **18** : 16-19.

Zade, V.R., Patel, V.N. and Zade, N.G. 1990. Standardization of seed testing procedures for *Trigonella foenum-gracum* and *Cyamopsis tetragonoloba*. *Ann. Plant Physiol.* **4** : 182-185.

Zheng, G.H., Gu, Zh and Xu, B.H. 1980. A physiological study of germination in guar (*Cyamopsis tetragonoloba*). *Acta Phytophysiol. Sinica* **6** : 115-126.

CHAPTER - 5

FORAGE ASPECTS

U.P. Singh and Dinesh K. Agarwal

Cluster bean (*Cyamopsis tetragonoloba* L. Taub) commonly known as Guar derived from Gujarati word "GAU AA HAR" meaning food of the cow is also known by many other name such as Khutti, Darari etc. In India guar crop is known as poor men's crop and has been grown from ancient times in western states of the country. It is an important dryland, drought hardy annual legume grown widely as multipurpose crop viz: forage, vegetable, grain (gum) and green manuring. Guar has acquired a great significance in recent years because of guar gum (galactomannan -- having a high molecular weight, is a carbohydrate polymer made up of a large number of mannose and glactose units linked together) contained in its endosperm, which is being exported in large quantities from India. Guar seeds are used for the extraction of gum and cattle feed. With the establishment guar gum and cattle feed factories, the demand of guar seed (grain) has increased significantly. Guar gum is exported to many countries mainly USA, UK, Japan, France, Italy and Netherlands etc. and it is treated as an important cash crop of arid regions.

Guar has also assumed important place as most dependable *kharif* fodder legume in many parts of north-west India, where rainfall is scanty and erratic. It also provides nutritious and protein rich guar meal for cattle and adds organic matters and also increases fertility to the soil by fixing considerable amount of atmospheric nitrogen. This crop does not require any specific *rhizobia*.

IMPORTANCE

As Fodder

Guar provides very palatable and nutritious fodder and guar meal (feed) to the animals. Being a leguminious crop, it is a rich source of protein but low in TDN values. On dry matter basis guar fodder contains about 16-20% crude protein, 46% TDN, 11-13% DCP and more than 70% digestibility, dry matter intake is quite high i.e., 2.50 to 3.55 kg/100 body weight. Nutritional value of fodder varies with the stage of harvesting, however, it is advisable to feed guar fodder at 50% flowering stage. Guar fodder can also be used as hay and silage.

As Guar meal

Guar meal or churi/chuni, a byproduct of gum industries consisting of seed coat and germ material, is good source of protein. It contains about 40-42% crude protein being one and half times more than available protein in guar seed. It is used as feed for livestock including poultry but has also deleterious factors such as residual guar gum and trypsin inhibitor. However, it can replace cent per cent protein supplement like groundnut cake in the ruminants. Chemical composition of guar meal is given below in Table 1.

Table 1. Average percentage composition of Gaur meal

Components	% range
Oil	05-06
Protein	35-45
Fiber	10-12
Carbohydrates	28-30
Mineral matter	05-06
Digestibility of protein	96-97

Source: Arora, S.K. (1981).

PRESENT STATUS OF GUAR IMPROVEMENT AS FORAGE CROP

Although Tropical Africa has maximum number of wild relatives of guar, but the genetic improvement in this crop has largely been brought about through selection from amongst the huge variability present in the land races collected from different parts of India (Singh, 1981). Guar is recognized as a multipurpose crop and has been subjected to extensive genetic modifications. Guar is a self pollinated crop bearing small flowers but do not present difficulties in crossing as its new crossing technique (Chaudhary *et al.*, 1974) made hybridization quite easier which may be explained as **"pressing the corolla (keel) by forceps and pulled out. In this process the anthers are safely pulled out by leaving gynoecium part intact without any injury which can be pollinated as per desire".** The inter-varietal variability is generally more conspicuous in respect of branching pattern ranging from single stemmed (unbranched) to profuse branching types with various intermediates forms varying in size of plant, leafiness branching pattern and pod bearing characteristics. The branched types generally produce small leaflets and pods and are cultivated as dry-land crops for grain purpose. The single stemmed types are generally tall and may be grown in mono-cropping system or mixed with other cereal crops in better rainfall situations.

Most of guar research was undertaken for either grain or fodder purpose in arid and semi-arid agroclimatic situations mainly at HAU, Hisar, RAU, Durgapura; GAU, Sardar Krushinagar, NBPGR, New Delhi, CAZRI, Jodhpur and IGFRI, Jhansi. However, the responsibility has been assigned mainly to IGFRI, Jhansi and HAU, Hisar to work on **forage aspects** of clusterbean. Intensive efforts were made to collect a large number of genetic diversity available within the country and abroad, which were further exploited through various breeding strategies. Most of the research and development aspects have already been attended by scientists of these institutions such as, new crossing technique (Chaudhary *et al.*, 1974), genotype X environment interaction for forage and grain aspects (Saini *et al.*, 1978, Paroda *et al.*, 1980 and Jhorar *et al.,* 1980), management and its extension (Saini *et al.*; 1979, Singh *et al.*, 1994, Singh and Dixit, 1994, Singh and Gupta, 1996). This way, HAU and IGFRI, Jhansi have been able to develop and release some of the forage cluster bean varieties viz., HFG-119, HFG – 156, HFG – 314 and Bundel Guar-1(IGFRI –212-1).

Later, priorities of users were reviewed and changed in light of the present situations and need for the development of dual type (forage cum grain type with high gum content) was felt. In view of this, institutions had planned their research programmes accordingly. Inter-varietal and inter-specific crosses have also been attempted but with limited success. The breeding programmes so far have mainly been focused on developing:

- faster growth rate with early maturing type,
- determinate fodder type with high leaf : stem (L/S) ratio,
- high palatabililty and nutrious,
- photo and thermo-insensitive types,
- disease resistant types and
- adequate seed with high gum and protein contents.

Besides, genotypes with better digestibility, high protein content and branched types may also be assessed for their fodder production potential. Generally the selection aims at high forage production good quality seeds. Fortunately, a number of improved varieties (forage, forage cum grain and grain type) have been developed at various organizations, certain notable varieties are : Malosan, Katchh, GG-1, FS-277, HFG-119, HFG-156, HFG-314, HG-258, HG-75, HG-182, HG-365, BG-1, BG-2, BG-3, DP-Safed, Durgajau, RGC-197, RGC-147, RGC-936, Durga Bahar, Naveen, PLG-85, IC-9229-3, Pusa Sadabahar, Pusa Mausmi and Pusa Navbahar, Guara-80 etc. (Table 2).

FUTURE THRUST FOR BREEDING FODDER CUM GRAIN VARIETIES

During future course major attributes required for development of fodder cum guar varieties are as follows:

- **Forage yield:** Fast growth and leafness
- **Forage quality:** High CP and in vitro dry matter digestibility (IVDMD), less fibrous stem and good palatability.
- **Photo-insensitivility:** For use as summer forage crop with assured seed production.
- **Synchronous flowering and pod maturity:** The optimal fodder and seed production.
- **Plant type :** Tall plants with upright branching to suit mix cropping with cereals.
- **Drought** and heat tolerance
- **Responsiveness** to fertilizers.
- **Free** from bacterial diseases and insect pest infestation.

Table 2. Varieties released at national or state level in guar for different purposes

Name of Institution	Fodder type	Fodder cum grain type	Grain type	Vegetable type
HAU, Hisar	HFG-119 HFG-156	FS-277	HG-258 HG-75 HG-182 HG-314 HG-365	—
IGFRI, Jhansi	Bundel Guar-1	Bundel Guar-2 Bundel Guar-3	–	–
GAU, S.K. Nagar	–	Malosan	Katchh-8 GG-1	–
NBPGR, New Delhi	—	—	Naveen, PLG-85, IC-9229-3	Pusa Sadabahar Pusa Mausmi Pusa Navbahar
RAU, Durgapura	—	—	DP Safe, Durgajai RGC-197 RGC-471 RGC-936 RGC-1002 RGC-1003 RGC-1017	Durga Bahar,
CAZRI, Jodhpur	—	—	Maru Guar	—
PAU, Ludhiana	Guara-80	—	—	—

VARIETAL DESCRIPTION

Details of the varieties released for fodder, fodder-cum-grain, grain type and vegetable types are given in Table 2.

HFG-119

This variety is tall, branched and leafy type with medium late maturity. It produces 320 q/h green fodder and 60 q/h dry fodder yield. Leaves are pubescent with serrated margin and the variety matures in 130-140 days producing 8-10 q/h seeds having about 30% gum content. The variety is moderately resistant to bacterial blight and powdery mildew.

HFG-156

This variety was released for fodder purpose for whole country. It is tall and branched with pubescent leaves with serrated leaf margins. It gives 261.4 q/h average green fodder yield in 70 days and matures in 120-140 days. It is also moderately resistant to bacterial blight and has good quality fodder.

Malosan

The variety is branched, leafy type with medium size leaves, medium maturity (115-125 days). It produces 250-300 q/h green fodder and 50-55 q dry fodder yield. It is slightly susceptible to Alternaria and bacterial blight disease under field conditions. The seeds are medium bold and whitish in color with a black tinge and produces 810 q/h seeds. In Gujarat this variety is being cultivated for forage cum grain purpose.

Bundel Guar-1

Bundel Guar-1 (IGFRI 212-1) is a rainfed fodder variety developed from an indigenous collection No. B-5/54 from Rajasthan. This variety is recommended for cultivation in all the guar growing areas of the country. It is most productive on well drained loamy to sandy loam soils and adapted to scanty rainfall areas in the tropical and sub tropical regions. Plants flower in 50-55 days and mature in 120-135 days. The values of per cent CP, ADF, NDG and IVDMD are 8-11, 43-49 and 72-74, respectively. The seed yield ranges from 0.5 – 0.6 t/h.

Bundel Guar-2

Bundel Guar-2 (IGFRI 2395-2) is a selection from indigenous collection 5496-2 from Jodhpur and is recommended for the entire guar growing areas as a rainfed fodder variety with substantial seed production potential. It is an erect, annual, fast growing variety, which attains the plant height of 90-120 cm. The green fodder and dry mater range from 25 to 30 t/h and 5 to 6 t/h, respectively. The IVDMD on dry matter basis is 74%.

Budel Guar-3

Bundel Guar-3 (IGFRI 1019-1) is the latest fodder guar cultivar released in 1999 as a dual type for guar growing areas of the country. The green fodder and dry matter yield for this variety range from 35-40 t/h and 5.5 to 6 t/h, respectively. The variety is tall growing (95-130 cm), erect, profusely branched type with intermediate growth habit and is resistant to major insect pests and diseases. Bundel Gaur-3 has high crude protein content (10.57 q/h) both in forage and seed as well. It has high value of digestibility (IVDMD 71.21%) in forages on dry matter basis. The variety is also resistant to lodging, free from pod shattering problem and highly responsive to fertilizers. Normal sown crop provides highest yield potential both for fodder and seeds (Fig. 1).

CULTURAL PRACTICES

Sowing time

The optimum sowing time for guar is first week of July. Fodder yield is understood to be reduced when planting is delayed to last week of July or first week of August. Guar for fodder as well as vegetable purposes during summer can also be sown from second week of March to first week of April.

Land preparation

No special seed bed preparation is required for this crop. Two to three ploughings are sufficient. The first ploughing may be done with turning plough and subsequent 1-2 ploughing done with Desi plough. Each ploughing should be followed by planking. At sowing, field must be well drained, with adequate moisture and should be free from the weeds.

Method of sowing and seed rate

The guar should be sown in rows 45 and 30 cm apart for branched and unbranched types, respectively. Thinning should be done after 15 to 20 days of sowing to maintain an optimum distance of 15 cm amongst the plants. The optimum seed rate of guar for fodder is 20 kg/h.

Crop rotations

Guar could be fitted well in most of the crop rotations. It is a deep, tap rooted summer/kharif season legume and is an excellent soil nourishing crop. It is suited well in rotation with wheat, barley, potato, cotton, bajra and sorghum. Guar significantly increases the yield of following crops by enhancing the soil fertility. It is found that rotations, of bajra-guar and bajra-mung proved to be more efficient with regard to total productivity per unit area.

Inter cropping

Under rainfed conditions, clusterbean is generally grown with pearl millet and sorghums as an intercrop. Investigations revealed that as a preceding crop clusterbean has beneficial effects on soil fertility as it fixes about 37-196 kg nitrogen/h annually in crop residues (Singh et al., 1994). Crop residues left in the field enrich the soil to increase the productivity of the subsequent crop. The results of a rotational study revealed that pearl millet – cluster bean gave 11% higher pearl millet yield as compared to continuous cultivation of pearl millet (Fig. 2 & 3).

Fertilization

Fertilizer may be applied to fertility status of the soil. Guar being a leguminous crop is capable of meeting its requirement of nitrogen through atmospheric nitrogen fixing process. However, a basal dose of 15-20 kg nitrogen/h can be given to boost up the initial growth. Guar, like most legumes, requires high levels of phosphorus. The 40 kg P_2O_5 /h has shown beneficial effect. On fertile lands or where preceding crops have been fertilized heavily, fertilizer for guar can be reduced or omitted. Application of 25 kg/h Zinc Sulphate in deficient souls has been observed profitable.

Weed control

Guar grown for fodder purpose does not require any intensive effort for weeding, although guar seed yield is reduced greatly by weed competition as weeds compete for moisture, nutrients and light. Periodical weeding and hoeing will keep the guar fields free from weeds. One to two hoeings may be done to check weeds and first hoeing should be done within 20-50 days of sowing. Basaline or Trifluarin @ 1.5 kg/h should be sprayed all over the field and incorporated in the upper 5 cm soil, prior to sowing.

Irrigation management

Cluster bean is generally grown as rainfed. However, during long duration of moisture stress, one or two supplemental irrigations, if available, can be applied. In summer, 2 to 3 irrigations are sufficient to meet transpirational requirements of the crop. During kharif season, proper drainage is more important.

Pests and diseases

Guar is susceptible to a number of diseases but bacterial blight and Alternaria leaf spot are the two major diseases causing serious damage to crop. Disease tolerant varieties and

Bundel guar 3 : grain cum fodder variety

Guar intercropped with sorghum

Guar intercropped with pearlmillet

disease free seed should be utilized and cultural practices that encourage the development of strong, healthy plants to reduce damage caused by disease may be followed as below :

1. **Bacterial leaf blight**: It is the most serious disease of guar. Bacterial blight is seed borne and may cause the death of plants at seedlings or at any other stages of growth. Bacterial leaf blight causes large angular lesions at the top of the leaves, which may lead to defoliation, and black streaking on stems. Seed treatment with hot water water at 56^0C for 10 minutes controls the disease.
2. **Alternaria leaf spot**: A fungus is the causal organism of this leaf disease. Symptoms vary from brown zonate to target like lesions on the leaf. Affected area may get enlarged and cause leaf dropping. The fungus becomes severe during the periods of heavy dew and high humidity. The disease can be controlled by spraying Dithane Z-78 (0.2%) and Cupramor (0.3%) at 15 days interval.
3. **Powdery mildew:** Powdery mildew is also caused by a fungus and manifests on the leaves only. The affected leaves are covered with white mycelial patches dotted with the fruiting bodies (perithecia). Severely infected plants get defoliated and weakened by pre mature drying up and death of infected leaves. The disease can be controlled by dusting any sulphur compound or spraying 0.01 per cent solution of Karathane.

SUMMARY

Guar provides palatable and nutritious fodder and meal to the animals, it is however, a poor source of T.D.P. Guar meal contains 40-42% crude protein. Guar meal is used as a feed for animals and poultry but has also deleterious factors such as guar gum and trypsin inhibitor. Genotypes with better digestibility, high protein content and branched types besides several other traits are preferred for fodder production. A number of genotypes improved in forage, forage-cum-grain and grain type have been developed. The important forage type varieties developed in recent past are HFG-119, HFG-156, Malosan, Bundel Guar-1, Bundel Guar-2, Bundel Guar-3. These varieties may produce almost 250-300 q ha^{-1} green fodder yield and 50-60 q dry fodder yield ha^{-1}. Improved cultural practices for guar fodder production have been developed, besides important insect pests and disease affecting fodder yield and quality have been identified and their control measured developed.

REFERENCES

Arora, S.K. 1981. Guar- Its uses are many and expanding all the times. *IVth Guar Research and Development Workshop* held at CAZRI, Jodhpur, March, 10-11, 1981.

Chaudhary, B.S., Paroda, R.S. and Solanki, K.R. 1974. A new crossing technique in cluster bean (*Cyamopsis tetragonoloba* L. Taub.). *Curr. Sci.* 43: 263-265.

Jhorar, B.S., Saini, M.L. and Solanki, K.R. 1980. Phenotypic stability of dry fodder yield in cluster bean. *Haryana Agric. Univ. J. Res.,* 10: 1-4.

Paroda, R.S., Saini, M.L. and Singh, J.V. 1980. Variety X environment interaction of cluster bean. *Forage Research*, 6: 71-74.

Saini, M.L., Jhorar, B.S. and Solanki, K.R. 1979. Genotype X environment interaction for green fodder yield in clusterbean. *Forage Research*, 5: 185-188.

Singh, M., Saxena, A. and Lodha, S. 1994. Grow clusterbean in arid and semi arid zones for vegetable and fodder. *Indian Farming*, 43(11): 28-30.

Singh, U.P. 1981. Genetic Improvement of Guar. *4th ICAR Guar Research and Development Workshop*, CAZRI, Jodhpur, April, 1981.

Singh, U.P. and Dixit, O.P. 1994. Bundel Guar-1. A cluster bean variety for high fodder production. *Indian Farming* 43(11): 31-33.

Singh, U.P. and Gupta, S. 1996. Grow Bundel Guar-2, for fodder and grain, *Indian Farming*, 46 (3): 15-16.

CHAPTER - 6

GENETIC RESOURCES

N.K. Dwivedi and Neelam Bhatnagar

INTRODUCTION

Clusterbean (*Cyamopsis tetragonoloba*) (L.) Taub. Syn. *C. psoralioides* (Lamk; D.C.), is a drought hardy cultivated arid legume crop of the Indian arid and semi-arid zones. The natural variability of this legume is available in drier regions of the country. Plant genetic resources of a particular crop are prerequisite for its continued genetic improvement, which are the heritage to be conserved for current and future use so as to ensure human welfare (Paroda, 1987). The present article deals with efforts made by NBPGR, Regional Station, Jodhpur in collection, introduction, characterisation, preliminary evaluation, documentation, maintenance, seed supply and conservation of variability, available in this legume.

ORIGIN OF CLUSTERBEAN

Genus *Cyamopsis* D.C. (Kyamos-bean and opsis-resemblance to) belonging to the tribe Galegeae of family-Papilionceae, has four species namely *C. dentata*, *C. senegalensis*, *C. serrata* and *C. tetragonoloba*. Senn (1938) suggested that *Cyamopsis* (n=7) is a derived aneuploid from the genus *Indigofera* (n=8). *C. senegalensis* and *Indigofera* resemble each other very remarkably. Gillete (1958), emphasized that *Cyamopsis* is a distinct genus with tropical Africa as its probably centre of origin. Chavalier (1939) postulated that *C. senegalensis* probably extended up to Sindh where after domestication a few of its cultigens became cultivated in India. However, Vavilov (1951) had suggested India as the geographical centre of variability for clusterbean, however, frequent occurrence of shattering type (an intermediate of the wild species and cultivated type) of clusterbean in Western Rajasthan (Dabas and Thomas, 1986) indicated that the clusterbean perhaps had domesticated in this region. Hymowitz (1972) believed that the African wild species *C. senegalensis* appeared to be the ancester of West African *C. tetragonoloba*.

GERMPLASM COLLECTION AND INTRODUCTION

A total of 4869 collections were made from different agro-ecological regions and farming systems by this station under region specific/crop specific/multi crop specific/individual/ collaborative germplasm collection missions. This collection comprised mainly of primitive cultivars and others under cultivation. Of 4869 accessions maintained at this station 737, 348, 181 and 1778 were collected from different parts of Gujarat, Haryana, Punjab and Rajasthan, respectively, during kharif seasons of 1980 to 1994. In all, 1816 accessions were received from NBPGR, Head quarters New Delhi. In addition, nine accessions were introduced from USA. Two wild species namely *C. senegalensis* and *C. serrata* were introduced from USA (Dabas and Phogat, 1993). Scientists from Central Arid Zone Research Institute have recently collected, nine collections from parts of North-west Rajasthan during October 1999 under NATP (Plant Biodiversity) programme (Annonymus,1999). Efforts have also been made at the State Agricultural Universities to strengthen the germplasm collections. There is still need to explore some pockets of Gujarat, Madhya Pradesh, Uttar Pradesh, Peninsular India, Pakistan and tropical Africa.

GERMPLASM PRELIMINARY EVALUATION

Morphological characterisation: Out of 4869 accessions, 3835 (9 exotic and 3826 indigenous) were evaluated. In general there was a wide range of variation in plant characters (Table-1). Studies on 520 accessions during kharif seasons of 1988-1989 and 1989-1990 exhibited that most collections from Rajasthan had early seedling emergence. About 57.8% of the collections had good initial vigour, whereas 35.2% had medium and 7% had poor vigour. Maximum initial vigour (63%) was observed in accessions collection from Rajasthan followed by those from Gujarat (20%) and Haryana (17%). Out of 3835 accessions characterized, 645 were siggle stemmed type and 521 smoothed leaf type. Foliaceous bracts were found in 24 collections from Mehsana district of Gujarat (Dwivedi *et al*, 1996). IC 39987 (Soila, Rajasthan) was the earliest to flower (26 days), whereas, IC 116804 (Sri ganganagar, Rajasthan) was earliest to mature and IC 10355 was latest to mature. The cluster length was maximum (22.00 cm) in accession IC 116914, collected from Mehsana district of Gujarat and minimum (2.27 cm) in IC 116592 collected from Jodhpur. The pods in single stemmed collections were large and fleshy. The brached type produced comparatively small and rough pods. In the present study, pod length was maximum (11.97 cm) in accession IC 116960 and minimum (3.36 cm) in IC 116798 (Bikaner).

Table 1. Range and mean of variability for different characters in clusterbean germplasm

Characters	Range		Mean
	Minimum	Maximum	
Days to 50% flowering	25.00	76.00	41.32
Days to 50% maturity	68.00	128.00	94.57
Plant height (cm)	15.00	238.00	67.23
Branches plant-1	00.00	29.00	4.77
Cluster plant-1	02.00	85.50	12.05
Cluster length (cm)	02.27	22.00	06.52
Pod length (cm)	01.60	11.97	5.42
Pods plant-1	03.70	412.00	41.77
Seeds pod-1	02.00	14.77	08.16
Pod weight plant-1 (g)	01.00	125.00	15.87
Seed weight plant-1 (g)	00.40	71.00	8.36
100 seed weight (g)	01.50	05.30	3.27
Gum content (%)	21.77	34.38	28.78
Protein content (%)	22.43	29.24	26.37

All the characterised accessions produced non shattering type pods. Accession IC 116565, collected from Barmer had the maximum number of branches (10.17), clusters (85.50), pods (198.33) and seed yield (41.76 g) per plant, while accession IC 116906 (Banaskantha, Gujarat) and IC 116522 (Jodhpur) followed in grain yield and number of clusters per plant. In some of the accessions there was a close relation among seed yield, number of branches, number of clusters, number of pods per plant and number of pods per cluster, as earlier observed by Arora and Gupta (1981); Sangi *et al*. (1964); Mital *et al* (1969) and Tripathi and Lal (1975). Out of 3835 accessions, 348 had more than 3.75 g 100-seed weight.

Variability was also recorded in size (small/medium/bold), shape (round/oval/rectangular/flat) and colour (grey/light grey/dark grey/pink/light pink/purplish grey/whitish grey/white) of seed. Most of the collections from Gujarat had smooth leaves and pink coloured seeds. Hymowitz and Matlock (1961) believed that the seed coat was environmentally controlled and was of no specific advantage. However, present investigators observed that seed colour was genetically controlled. Commercially seed size and shape are more important than colour. The clusterbean varieties with 100 seed weight of 3.0 g onwards with oval shape are preferred by the guar gum industries due probably to easiness in processing (Dabas, 1993).

Quality characters: Gum and protein contents were estimated (Figs. 1 and 2) in 520 clusterbean accessions (Dwivedi *et al*, 1999). Gum content varied considerably and ranged from 19.1 to 34.1% (Menon *et al*., 1970), 14.6 to 15.5% (Dass *et al*., 1973), 11.23 to 26.33% (Tripathi and Srivastva, 1975), 22.4 to 42.6% (Paroda *et al*., 1977), 15.9 to 31.8% (Singh *et al*., 1977 and Dabas, 1993). Similarly, protein content in guar germplasm varied from 26.85 to 31.00% (Essar, 1957; Casares and Herrera, 1960; Misra *et al.,* 1968).

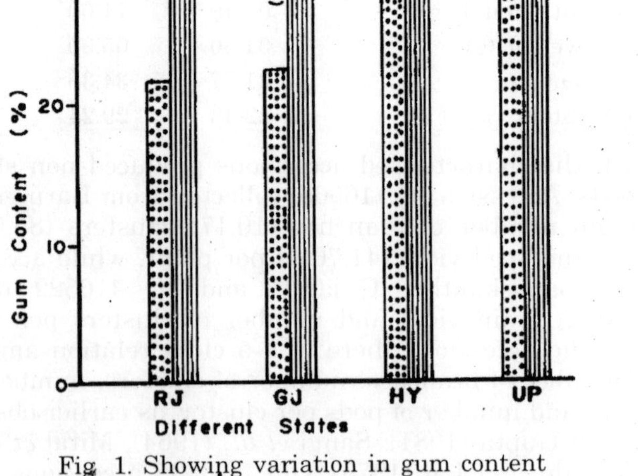

Fig 1. Showing variation in gum content
(values in parenthesis are mean Gum content)

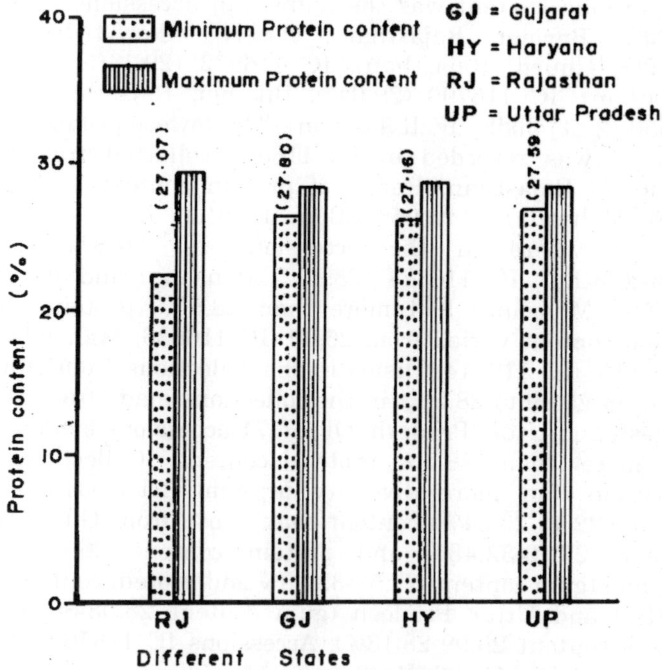

Fig. 1. Showing variation in gum content (values in parenthesis are mean Gum content)

The highest gum content was recorded in accession IC 116601 (34.38% collected from Bikaner, Rajasthan), closely followed by accession IC 16752 (34.07%, collected from Nagour, Rajasthan) and IC 116682 (33.91%, collected from Churu, Rajasthan). Accessions IC 11601 and IC 116682 were single stemmed. Most of the lines collected from Bikaner, Churu and Nagaur districts of Rajasthan possessed comparatively higher gum content, followed by collected from Jaisalmer and Barmer districts. Gum content was maximum in IC 116906 (32.48%) collected from Banaskantha and minimum in IC 116926, (27.19%) collected from Palanpur, while it was maximum and minimum in accessions IC 116817 (32.18%, Mahendragarh) and IC 116814 (27.53%, Hisar) collected from Haryana, respectively. Of the 71 accessions collected form Gujarat and 28 from Haryana; 22 and 6 found to have more than 31% gum content from the respective states. In the seeds collected from Madhya Pradesh and Uttar Pradesh, the gum content ranged from 28.03 to 30.6%. From a total of 520 accessions, 44 had more than 32% gum content.

Protein content was the highest in accessions IC 116558 (29.24%, Barmer, Rajasthan), closely followed by 116732 (29.19%, Churu, Rajasthan), IC 116572 (29.14%, Jaisalmer, Rajasthan), IC 116790 (29.04%, Bikaner, Rajasthan) and IC 116550 (29%, Jodhpur, Rajasthan). The lowest protein content (22.43%) was recorded in IC 116659 collected from Churu district of Rajasthan. Range of protein content varied from 25.86 (Mehsana) to 28.95% (Banaskantha), in collection from Gujarat, where in the accessions IC 116877 (28.58%, Banaskantha), IC 116884 (28.95% Mehsana) and IC 116886 (28.55%, Mehsana) had more than 28.50% protein content. Protein content varied from 26.06 (IC 116836, Mahendragarh) to 28.81% (IC 116814, Hisar) in the collections from Haryana, and from 26.99 to 28.53% in the collections made from Madhya Pradesh and Uttar Pradesh. Of the 71 accessions evaluated, 31 had more than 28.5% protein content. Collections from Rajasthan had more diversity in gum (21.77-34.38%) and protein (22.43-29.24%) content than those from Gujarat (gum content, 27.19-32.48% and protein content 26.19-28.95%), Haryana (gum content, 27.53-32.18% and protein content 26.06-28.81%) and Uttar Pradesh (gum content 28.03-30.60% and Protein content 26.99-28.13%). Accessions IC 116568 (Barmer, Rajasthan), IC 116906 (Banaskanatha, Gujarat) and IC 116522 (Jodhpur, Rajasthan) had high grain yield per plant (41.78, 31.60 and 30.95 g, respectively) with high gum content (30.40, 32.48 and 31.62% respectively). Menon *et al.* (1970) and Mital *et al.* (1971) also reported positive correlations between grain yield and gum percentage.

Multilocational trials: Two hundred accessions were initially evaluated at four different locations, of Central Arid Zone Research Institute, Jodhpur; Gujarat Agricultural University S K Nagar; CCS, HAU, Hisar and Rajasthan Agricultural University, ARS, Jaipur. During the course of study, early maturing and high yielding accessions were identified for different locations. Accession guar-113 proved higher yielder at Gujarat and Jodhpur, while collection CH 49-4 gave superior yield at Jaipur and Hisar locations.

PROMISING ACCESSIONS IDENTIFIED WITH SPECIFIC TRAITS

Early maturity: IC 116804,- 116868, - 116869,-116930, HG1-1/P8-8-1-1-1, HG7-3-2-1-1, HG23-3/P1-1,

High yield:	IC 116522, -116546,-116565,-116568, 116906.
High gum content:	IC 116577,-116601,-116609,-116627.-116676,-116682,-116752

GERMPLASM DOCUMENTATION

Based on the variability study, 4 catalogues (Chopra *et al.*, 1983, 1985; Dabas *et al.,* 1989 and Dwivedi *et al.,* 1995) have been published documenting the collected information for dissemination amongst the users.

GERMPLASM CONSERVATION

Conservation refers to the management of natural resources for the suitable benefits of present and to meet the need and aspiration for the future generations in all 609 accessions have been deposited in National Gene Bank at NBPGR, New Delhi for long term conservation. In addition, 472 accessions have been processed for the Medium Term Storage Facility available at this Station.

GERMLPASM SUPPY

Approximately 2115 different accessions were supplied to the various indentors on their demand.

FUTURE THRUST

1. Collection of germplasm from unexplored areas like Kuchchh and other non-traditional parts.
2. Introduction of more exotic germplasm.
3. Detailed evaluation specially for vegetable, fodder and disease resistant traits
4. Development of core collection/multilocational trials.
5. Analysis of all the available accessions for gum and protein content.
6. Identification of accessions with possessing desired traits at a time for users.

SUMMARY

A total of 4869 collections were made from different

agro-ecological regions and farming systems. Two wild species (*Cyamopsis senegalensis* and *C. serreta*) were introduced from USA. Out of 4869 accessions 3835 (9 exotic and 3826 indigenous) were evaluated. Foliaceous bracts were found in 24 collections made from Mehsana district of Gujarat. Accessions IC 116565 (from Barmer) and IC 116601 (from Bikaner) had maximum seed yield per plant (41.76 g) and gum content (34.07%) respectively. Two hundred accessions were initially evaluated at four different locations. Promising accessions were identified for their early maturity (7 accessions), high yield (5 accessions) and high gum content (7 accessions). Four catalogues have been published. 609 accessions have been deposited in National Gene Bank at NBPGR, New Delhi for long term conservation and 472 accessions have been processed for the Medium Term Storage Facility at this station. 2115 different accessions were supplied to the various indentors. Work on future thrust are suggested.

REFERENCES

Anonymous, 1999. *Annual Report,* NBPGR Pusa Campus, New Delhi.

Arora, S.S. and Gupta, B.S. 1981. Genetic variability, correlation and path coefficient analysis in guar [*Cyamopsis tetragonoloba* (L.) Taub.]. *Curr. Agric.* **5(3-4)**: 116-122.

Casares, R. and Herrera, C.L. 1960. Estudio bromatologico del guar, *Stizolobium decringianum Y. Phaseolus radiatus, nucvas leguminosas aclimatadas en Espana* (Bromatological study of guar). Stizolobium decringianum and *Phaseolus radiatus*, new legumes and that have become acclimatized in Spain 7. An. Bromatol **12**: 405-410.

Chevalier, A., 1939. Research on species of the genus *Cyamopsis* forage plants for topical and semi arid countries. *Rev. Bot. Appl.*, **19**: 242-249.

Chopra, D.P., Thomas, T.A. and Singh, R., 1985. *Catalogue on guar germplasm* N.B.P.G.R. Regional Station, Jodhpur, 200p.

Chopra, D.P., Thomas, T.A., Mehra, K.L. and Singh, R., 1983. *Catalogue on guar germplasm* N.B.P.G.R. Regional Station, Jodhpur, 55p.

Dabas, B.S., 1993. Conservation on genetic resources of arid legumes. In : *Conservation and Management of Plant Genetic Resources.* pp. 164-170 (Eds. Rana, R.S., Saxena, R.K., Saxena, S. and Mitter, V.),. National Bureau of Plant Genetic Resources, New Delhi.

Dabas.B.S. and Phogat, 1995. Genetic Resources of arid legumes in

India. *Indian J.Pl. Genet.Resour.* **8**: 261-263.

Dabas, B.S. and Thomas, T.A. 1986. Shattering guar and its significance. *Inter. J. Trop. Agric.* **2**: 185-187.

Dabas, B.S. and Thomas, T.A., Rajni , N. and Chopra, D.P., 1989. *Catelogue on guar [Cyamopsis tetragonoloba* (L.) Taub.] germplasm. N.B.P.G.R., New Delhi 131p.

Dass, S.N., Arora, N.D. and Gupta V.P., 1973. Heritability estimates and genetic advance for gum and protein content along with seed yield and its components in clusterbean [*Cyamopsis tetragonoloba* (L.) Taub.] *Haryana Agri. Univ. J. of Resour.* **3**: 14-19.

Dwivedi, N.K., Bhandari, D.C., Dabas, B.S., Agarwal, R.C., Mandal, S. and Rana, R.S., 1995. *Catalogue on clusterbean* [*Cyamopsis tetragonoloba* (L.) Taub.] germplasm. N.B.P.G.R., New Delhi, 50p.

Dwivedi, N.K., Bhandari, D.C., Dabas, B.S., Mandal, S. and Rana, R.S. 1996. Variability in clusterbean germplasm with Foliaceous Bracts. *Annals Arid Zone* **35** : 151-153.

Dwivedi, N.K., Bhandari, D.C., Bhatnagar, N. and Dabas, B.S., 1999. Characterizing clusterbean [*Cyamopsis tetragonoloba* (L.) Taub.] germplasm for yield component and quality traits. *Ann. Arid Zone*, **38 (2)**: 151-156.

Essar, J., 1957. Guar- a versatile crop. *Seed World.* **80**: 12-15.

Hymowitz, T., 1972. The Transdomestication concept as applied to guar. *Econ. Bot.* **26**: 49-60.

Menon, V., Dubey, M.M. and Bhargava, P.D., 1970. Gum content variations in guar [*Cyamopsis tetragonoloba* (L.) Taub.]. *Indian J. Hered.* **2**: 55-58.

Misra, D.K., Bhan, S. and Prasad, R., 1968. Guara – a multipurpose summer legume. *Allahabad Fmg.* **42**: 239-244.

Mittal, S.P., Swarup, V., Kohli, M.M. and Singh, H.B., 1969. Variability in guar. *Indian J. Genet.* **29**: 98-103.

Paroda, R.S. and Saini, M.L., 1978. *Guar breeding. Forage Res.* **4A**: 9-39.

Paroda, R.S., Saini, M.L. and Jhorar, B.S., 1977. Genetic improvement of guar-*problems and perspectives. Workshop on Guar Research and Production* held at Central Arid Zone Research Institute, Jodhpur on January 11-12. 24-47.

Sanghi, A.K., Bhatnagar, M.P. and Sharma, S.K., 1964. Genotypic and phenotypic variability in yield and other quantitative characters in guar. *Indian J. Pl. Breed.* **24**: 164-167.

Senn, H.A., 1938. Chromosome number relationships in the leguminosae. *Bib. Gen.* **12**: 175-336.

Tripathi, R.M. and Lal, S., 1975. Estimate of variability and heritability of some quantitative traits in guar [*Cyamopsis tetragonoloba* (L.) Taub.]. *Indian J. Farm Sci.* **3**: 28-31.

Tripathi, R.M. and Srivastava, G.P., 1975. Note on the gum content of some new stains of guar (*Cyamopsis psoroliodes*). *Indian J. Agric. Res.* **9**: 153-154.

Vavilov, N.I., 1950. The origin, variation, immunity and breeding of cultivated plants. *Chronica Botanica* **3**: 364.

CHAPTER - 7

MICROBIOLOGY

A.V. Rao

Emerging economic importance and dependence of large human populations on arid agriculture, have resulted in increased biological and environmental research in the arid and semi-arid areas particularly during the last 25 years. The arid areas experience an annual rainfall of 100-400 mm with a coefficient of variation of 40-70 per cent. Low and erratic distribution of rainfall coupled with extreme temperatures result in frequent crop failures and considerably affect the agricultural economy in the region. Occurrence of consecutive droughts leads to wind erosion and creeping of sand dunes covering fertile agricultural lands. For many years, arid areas have been considered economically unimportant. During the past three decades, however, the economic agricultural utilization of arid lands had emerged as a critical element in maintaining and improving the Country's food supply. Incidence of intense solar radiation coupled with high wind velocity led to high evapo-transpiration – a situation most intriguing to arid lands.

The arid soils are broadly referred to as aridisols (Sierozems), with Camborthids and Solorthids as the major groups. These soils are extremely light, sandy, loamy sands or sandy loams. These are susceptible to wind erosion. Because of low contents of clay and silt the moisture holding capacity is low resulting in high evapo-transpiration rates. Subsoil salinity is very common probably due to extreme aridity. These soils are poor in organic matter content and other nutrients. In general, plant productivity in arid regions is very poor because of the limited natural resources and harsh climate. To achieve higher plant productivity, nitrogen can be provided either through chemical fertilization or biological nitrogen fixation because of low fertility status of arid soils (Table 1).

Table 1. Characteristics of Indian arid soil

Parameter	Quantity
Soil classification	Hypothermic typic haplocambids
Sand (%)	87.1
Silt (%)	5.5
Clay (%)	6.9
pH (1:2.5)	8.1
Electrical conductivity (dSm^{-1})	0.2
Organic matter (%)	0.34
Total N (mg kg^{-1})	310
Organic P (mg kg^{-1})	50
Olsen P (mg kg^{-1})	7.0

Bio-inoculants containing biological nitrogen fixing micro-organisms play a significant role in enhancing the productivity of arid legumes. Further, both economic and environmental considerations strongly favour biological nitrogen fixation as the process of choice in world agriculture especially in India. Biological nitrogen fixation requiring a relatively simple and localized technology and using mostly renewable energy sources has an important role to play in the immediate future. The annual turnover of biological nitrogen fixation from different sources on this globe is estimated to be 172 million tonnes per year. The amount of chemically fixed nitrogen in industries is about 2.5 to 4 times less than that of the biologically fixed nitrogen. Further biological nitrogen fixation through legumes is equivalent to or atleast half that of the industrial fixation .West (1975) estimated the total biological nitrogen fixation in arid zones of this globe to be 1.15 x 10^6 t N yr^{-1} and the average rate of fixation as 3.5 kg N ha^{-1} yr^{-1}, since the process is restricted to a small wet period. Among various biological nitrogen fixing systems, legume-*Rhizobium* symbiosis predominantly contributes to nitrogen enrichment of arid soils.

Nodulation of Clusterbean under Natural Conditions

Symbotic nitrogen fixation through legume-*Rhizobium* association is the most popular in any ecosystem. Clusterbean is a typical legume grown in tropical arid areas extensively. However, its production is low due to harsh environment besides low fertility status of sandy loam soils. In aridisols nodulation of clusterbean is poor and sometimes absent due to lack of optimum soil moisture, and sufficient population of *Rhizobium* in the soil which varies from zero to 10^4 g^{-1} soil (Anonymous, 1980). In general, nodulation is influenced by factors like soil type, pH, organic matter content, rhizobial populations in the soil, nature of the host plant etc. (Beadle and Tchan, 1953; Brockwell, 1976; Rao and Venkateswarlu, 1983).

Like other legumes, significant differences in genetic variability with respect to nodulation and N_2-ase activity exist in clusterbean (Rao, *et al.,* 1984; Khurana *et al.,* 1978). However, no correlation was found between nodule number or nodule weight and N_2-ase activity. Further, Khurana *et al.*, (1978) could not observe any relation between nodulation and grain yield in different varieties. Oke (1967) reported maximum N_2-fixation in non-branching types as compared to branching ones. Some varieties of clusterbean displayed poor nodulation and N_2-ase activity, despite being high yielders. Nodulation pattern and N_2-fixing ability of a variety at any given time may not show any relation with total N_2-fixation and consequent contribution to the given yield. It is evident that selection for efficient nitrogen fixing strains should be followed by hybridization to evolve varieties with both high nitrogen fixation and seed yield.

In general, there is a diurnal variation in N_2-ase activity in legumes due to variation in the supply of photosynthate to the site of N_2-fixation but Swaraj *et al.* (1988) reported that there was no such diurnal variation in the rate of N_2-fixation measured at 2 h intervals over a period of 24 hours in clusterbean. Venkateswarlu *et al.* (1983) observed maximum N_2-ase activity at 30 days of sowing and a decline thereafter. Lynd and Ansman (1989) observed that highest nodule development at plant density of 25 plants m^{-2} irrespective of the fertility status. Nodulation was significantly correlated with shoot growth. However, N_2-ase activity $plant^{-1}$ declined significantly with plant maturity (Oke, 1967).

Microsymbiont

Richmond (1926) found that *Rhizobium* from cowpea could produce nodules on clusterbean and pure cultures obtained from these again formed nodules on cowpea roots. Further, Raju

(1936) observed nodulation in clusterbean by the strains of *Rhizobium* isolated from different hosts of cowpea group. These observations formed the basis for placing clusterbean in cowpea cross inoculation group.

Rhizobial strains isolated from the clusterbean nodules are mostly slow growing and produce profuse polysaccharides. Garg *et al.* (1993) reported that the strain (CT-2014) isolated from clusterbean was to nodulate different heterologous hosts and it was at par with that of the homologous strain. They also isolated some strains which were found to grow on various media containing succinate, fumerate, malate, mannitol and gluconate. These strains vary in their requirements for induction of *ex-planta* N_2-ase activity depending on the operational pathway of C-utilization. In another study, Dadarwal *et al.* (1985) observed that about 50% of the rhizobial strains of summer legumes including clusterbean are Hup^+. In summer legumes, inoculation with both Hup^+ and Hup^- strains showed that the plants were predominantly nodulated by Hup^+ strains indicating thereby, the competitiveness of Hup^+ strains. Ahlawat *et al.* (1992) reported that bacteriocin producing character of the rhizobial strains was found to compete effectively with other strains and could be used as a marker in various ecological studies.

Legume-*Rhizobium* interaction:

Erdman (1948) observed variation in nodulation and nitrogen fixation in clusterbean among different strains of *Rhizobium*. This crop had shown a significant response to inoculation in terms of nodulation and N_2-fixation (Venkateswarlu *et al.* 1982). About 8- fold increase in N_2-ase activity was recorded in the inoculated plants and there is a strain-variety interaction in clusterbean (Fig. 1). Some strains are more compatible with all the varieties as compared to other strains. Stafford and Lewis (1980) observe 30% more nodules in the inoculated plants compared to that of the uninoculated plants. Further, inoculated plants are taller with more nodules on the main stem and weighed more than that of the uninoculated plants. Though there was an improvement in nodulation upon seed inoculation, loamy sand soil supported better nodulation compared to that observed in sandy loam soil (Mand *et al.* 1991), however, nodulation was similar with both fast and slow growing strains.

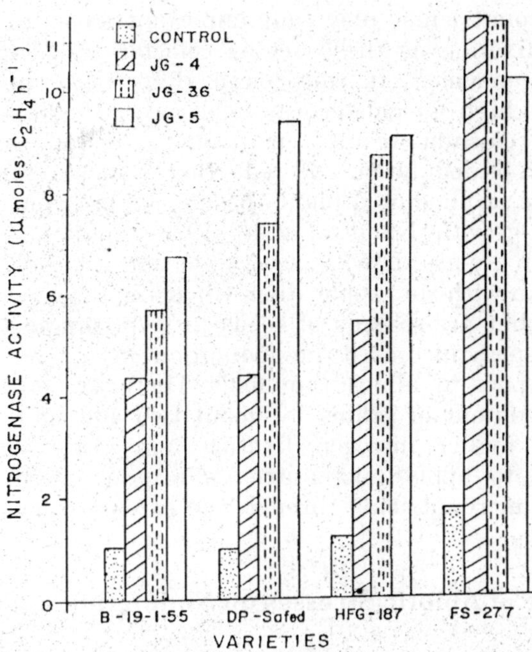

Fig. 1. Nitrogenase activity as influenced by variety-strain interaction

A significant increase in seed yield and dry matter production was recorded after seed inoculation with efficient strains of *Rhizobium* (Table 2). Increase in yield varied from 8-15% depending upon the intensity and distribution of rainfall. Clusterbean fixes about 30-70 kg N ha^{-1} and leaves a residual effect equivalent to about 15-20 kg N ha^{-1} (Rao, 1995). Phosphorus application was found to improve the efficiency of *Rhizobium* (Raut and Ali, 1983, Wani and Patil, 1982, Lynd and Ansman, 1989).

Table 2. Effect of rhizobial inoculation on clusterbean*

Parameter	Control	Inoculated
Nodule No. $plant^{-1}$	6.2	20.7
N_2-ase activity*	0.89	3.72
Grain yield (kg ha^{-1})	457	523 (14.4)
Straw yield (kg ha^{-1})	959	1180 (23.2)

(Average of 3 years)

μmoles C_2H_4 $Plant^{-1}$ h^{-1}

Figures in parenthesis indicate % increase over control.

Micronutrients play an important role in biological nitrogen fixation as these act as cofactors or components of N_2-fixing enzymes. It was found that the requirement of calcium is high for clusterbean (Bell *et al.* 1989). Sub-optimal levels of calcium results in delaying nodule appearance, prevention of nodulation and reduction in nodule numbers and percentage of noudlated plants beside, suppression of the root and shoot growth. Nandwal *et al.* (1990) reported a significant increase in N_2-ase activity, seed yield and nitrogen content of clusterbean with zinc application. Ghonsikar and Saxena (1993) reported that pre-soaking of seeds in tryptophane followed by 0.2% Mo spray increased the nodulation and enhanced the dry weight of pods by almost four folds. There was an increase in nitrogen content of grains by about two and half times. For achieving higher nitrogen fixation and grain yields, it is necessary to apply micronutrients directly or improve the status of micronutrients through the extensive use of organic manures.

Influence of Abiotic Stresses on Nodulation and Nitrogen Fixation

Legumes are normally cultivated on marginal lands on areas experiencing one or the other abiotic stresses such as salt, temperature, water and nutrient stresses. Clusterbean is generally recommended for arid or semi-arid areas where these abiotic stresses are very common resulting in lower productivity of clusterbean. Effects of various stresses on nodulation and nitrogen fixation in clusterbean are therefore discussed below.

Salt Stress

Clusterbean is reported moderately tolerant to sensitive for salinity but there is a great variation in salt tolerance amongst different genotypes (Lahiri *et al.* 1987; Francois *et al.*, 1990). The reduction in seed yield was mainly attributed to the reduction in photosynthesis, biological nitrogen fixation and other metabolic activities (Sprent, 1976; Rao and Venkateswarlu, 1987; Arayangkool *et al.*, 1990).

Salinity delays nodule formation, decreases number and weight of nodules and inhibits nitrogen fixation (Garg *et al.*, 1984; Kumar, 1987). Nodulation was progressively inhibited in clusterbean with increasing salinity or sodicity. Garg *et al.* (1986) reported that among different salts $NaHCO_3$ was the most detrimental resulting in 47% decrease in nodule number and 54% decrease in nodule weight. They further found a progressive reduction in nodule number and nodule weight with increasing salt concentration of irrigation water applied at the flowering stage. At high salt concentration (120 me L^{-1}), there was a significant reduction in number and weight of nodules but N_2-ase activity could not be detected (Table 3).

Table 3. Interaction of water and salinity stress on nodulation and N_2-ase activity

Salt conc. (me L^{-1})	Nodule number (plant $^{-1}$)		Nodule dry weight (mg plant $^{-1}$)		N_2-ase activity (n moles C_2H_4)	
	With stress	Without stress	With stress	Without stress	With stress	Without stress
0	17.5	30.5	21.5	75.0	177.4	499.5
40	16.0	25.0	16.5	59.0	108.2	458.7
80	12.5	16.5	–	46.0	26.0	270.2
120	10.5	12.0	13.7	18.0	0.0	0.0

(Garg *et al.*, 1986)

The reduction in nodulation might be due to the inhibitory effect of salts on the multiplication of rhizobia in the rhizosphere, availability of few root hairs for infection and/or hampering the root hair infection process. In general *Rhizobium* strains can survive and grow at salt concentrations that are inhibitory to most of the legumes but there was difference amongst the species and strains of *Rhizobium* with respect to salt tolerance to different salts (Bhardwaj, 1975; Elsheikh and Wood, 1990). Yadav and Vyas (1971) reported that clusterbean *Rhizobium* was found to be stable even at 3% salt concentration. In spite of optimum growth of *Rhizobium*, increasing levels of salinity and alkalinity in the rooting medium resulted in devoid of root hairs, mucilaginous layer and infection thread formation in the root system (Lakshmikumari *et al.*, 1974; Tu, 1981).

Nitrogen fixation in nodules mostly depends on the host plant for the supply of photosynthate, nutrients and water (Balasubramanian and Sinha, 1976). During salinity stress, rate of photysynthesis is reduced resulting in reduced supply of photosynthates to the site of nitrogen fixation in nodules. A significant reduction in N_2-ase activity with various sodium salts was observed through reduction in nodule number and weight (Garg *et al.*, 1984). Among the salts, $NaHCO_3$ was more detrimental, but the specific N_2-ase activity enhanced at lower concentrations. Sprent (1984) observed that at low salt levels, legumes may compensate by producing larger nodules.

Generally nitrogen fertilized legumes are more tolerant to salt stress as compared to those dependent on nitrogen fixation only (Bernstein and Ogata, 1966) and this was attributed that the fertilizer nitrogen alleviates the inhibitory effects of NaCl and Na_2SO_4 on N_2-ase activity. Similarly increasing external calcium concentration have been shown to inhibit Na absorption in various legumes and ameliorated the adverse effects of various salts on nodulation and N_2-fixation (Akhavan-Kharazian *et al.*, 1991). Calcium supplementation as $CaCl_2$/ $CaSO_4$ significantly ameliorated the adverse effects of NaCl on various enzymes involved in nitrogen metabolism in clusterbean and this was reported to be due to the maintenance of selective permeability of membrane resulting in the higher uptake of Ca and K and reduced Na uptake (Garg *et al.*, 1997).

Temperature Stress

In general nodule nitrogenase functions over a temperature range of 6-40°C with the maximum activity between 24-33°C and little at higher temperatures. Plants respond to lower root temperature by producing a larger volume of nodule tissue and this may be regarded as a form of compensation against lower nitrogen fixing ability per unit of tissue (Dart *et al.*, 1976). Arayangkool *et al.* (1990) reported a reduction of N_2-fixation by 80% and nodule weight by 50% at 40°C in clusterbean. They further observed significant interactions between many factors such as variety, strain and plant nutrition with temperature in respect to nodulation, nitrogen fixation and dry matter production. Leghaemoglobin synthesis in nodule is low at high temperature (Dart *el al.*, 1976), but when the plants experience optimal temperature the recovery of biological nitrogen fixation is fast. In general the plants provided with the starter nitrogen dose were not affected by temperature in clusterbean and maintained higher nitrogen fixation and dry matter production.

Light Stress

It is common to grow clusterbean in agro-forestry systems where light intensity is reduced depending upon the tree canopy. The shading by the tree species is a major factor affecting nodulation and nitrogen fixation (Sprent, 1973). Swaraj *et al.* (1988) studied the effect of darkness on clusterbean and it was found that exposure of 6 week old plants to darkness brought a rapid decline in N_2-fixation with acetylene reduction showing a 54% decrease after 18 h and 80% after 66 h of exposure to darkness. It was attributed to the accompanied decrease in leghaemoglobin content by 12-21 % and total soluble protein in nodule by 11-15%. Bhupinder Singh *et al.* (1994) observed an increase in amide content in nodules under darkness indicating the inhibition of transport of fixed nitrogen to the shoots. Further, transferring the plants from dark chamber to natural light for 10 hours partially restores the nitrogen fixing activity of nodules. On contrary, there was significant improvement in the levels of chlorophyll, protein and amino acids and, activities of various enzymes such as NR, GDH and GS in clusterbean under 25% shade compared to those without shade resulting in higher dry matter production and grain yields (Vyas *et al.,* 1996).

Water Stress

Biological nitrogen fixation plays an important role in meeting the nitrogen requirements of grain legumes and also in the enrichment of soil nitrogen status for use by the succeeding cereal crops. However, in arid and semi-arid regions, frequent droughts not only limit the growth and yield of cultivated grain legumes but also restrict their nitrogen fixation (Sprent, 1971; Venkateswarlu *et al.*, 1983). A number of studies (Kuo and Boersma, 1971; Engin and Sprent, 1973; Kaur *et al.*, 1985) both on tropical and temperate legumes have shown that water stress affects nodule initiation, structure, development and also nitrogenase activity.

Withholding of water led to a rapid decrease in the plant water potential in clusterbean (Rao and Venkateswarlu, 1987). The nodule number was not affected markedly under mild to moderate stress (a drop of about – 0.5 M Pa). Severe stress caused a significant reduction in nodule number (Table 4). Nodule fresh weight showed a significant decrease due to water stress. The decline in nodule fresh weight is due to both dehydration, cessation of growth and shedding of nodules.

Table 4. Nodulation and nitrogenase activity of clusterbean as influenced by water deprivation.

Treatments*	Plant water Potential (M Pa)	Nodule No. Plant^{-1}	Fresh wt. of nodules (mg plant^{-1})	N_2-ase activity**	Maximum activity retained (%)
Control (Do)	-1.20	14.8	113.5	509.0	–
D2	-1.40	15.5	98.2 (13.4)	500.0	98.2
D3	-1.60	15.4	74.5 (34.4)	319.9	62.8
D4	-1.90	13.0	54.3 (52.1)	98.2	19.3
D5	-2.20	12.3	41.2 (63.7)	73.5	14.4
D6	-2.40	9.5	24.5 (78.4)	20.4	4.0
D7	-2.70	8.2	22.4 (80.3)	Nil	–
LSD (p=0.05)		NS		54.5	–

* Number of days of with-holding water

** η moles C_2H_4 plant^{-1}; Figures in parenthesis indicate % decrease over control.

With decreasing plant water potential due to water deprivation, there was a sharp reduction in N_2-ase activity. With only a stress level of –0.4 M Pa, the reduction in N_2-ase activity was lowest in clusterbean compared to mung bean and moth bean. It may be noted that in clusterbean a higher percentage of its maximum activity was retained at all levels of stress compared to the other legumes (Fig 2). Further *in-vitro* studies indicated that the loss of nitrogenase activity is directly proportional to the loss of nodule moisture. But the per cent loss in N_2-ase activity at a given per cent loss in nodule moisture is more with mung bean nodules compared to that of clusterbean indicating that nitrogen fixing activity in clusterbean can function even under drought conditions. With 10% loss in nodule moisture, there was a reduction in N_2-ase activity by 48% and 17% respectively in mung bean and clusterbean respectively.

The fall in N_2-ase activity was higher and sudden at the vegetative stage compared to the flowering stage at a comparative degree of water stress (Venkateswarlu *et al.*, 1983). In clusterbean water stress during the vegetative stage may be more detrimental for the total nitrogen fixed. The reduction in N_2-ase activity in stressed plants may be due to inhibition of respiratory activity of the nodules affecting the ATP supply (Sprent, 1976) and non-availability of photosynthates (Huang *et al.*, 1975).

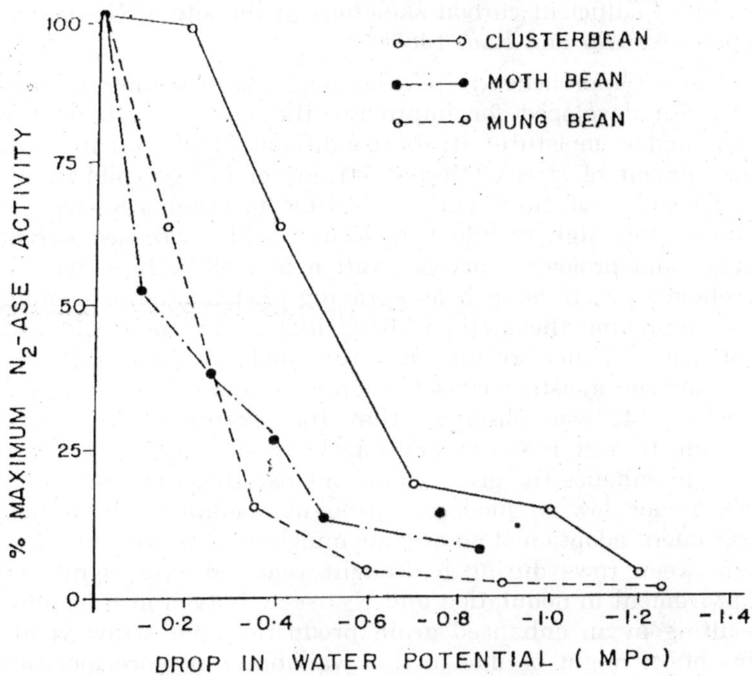

Fig. 2. Relationship between nitrogenese activity and drop in plant water potential.

Upon rewaterig the droughted plants, the recovery of N_2-ase activity was maximum and rapid in clusterbean compared to the other legumes. It has generally been found in several legumes the inhibition of N_2-ase activity becomes irreversible if the water content of nodules declines below 80% (Sprent, 1976). However, in clusterbean the stress mediated depression in acetylene reduction was found to be reversible even when the nodule moisture content dropped to 60 to 70% at the vegetative and flowering stages, respectively. This could be due to the greater drought tolerance compared to the other legumes. Further meristematic type of nodules found in clusterbean was reported to have a greater recovery capacity than the spherical type (Engin and Sprent, 1973). This seems to be related to the initiation of fresh growth in meristematic nodules upon rewatering. In view of the above, N_2-fixation process in clusterbean is highly sensitive to soil moisture stress but at the same time it has a higher capacity to recover under favourable moisture conditions. With increasing stress intensity there was an accumulation of ureides in nodules and it was due to lack of sufficient carbon skeletons at the site of N_2-fixation (Venkateswarlu and Rao, 1987).

Some approaches to alleviate the stress mediated effects had been developed for improving the symbiotic N_2-fixation even under moisture stress conditions. These include (a) development of stress tolerant strains of micro-symbiont, (b) identification of host variety having meristematic type of nodules and high nodule turgiscence, with increased carbon supply and prolonged active fixation and (c) manipulation of agronomic practices such as spraying photysynthetic stimulators, improving the method of inoculation and pelleting with synthetic polymer resins/ moisture holding materials and adopting soil moisture conservation techniques such as organic mulching. It was observed that the strains of *Rhizobium* tolerant to water stress developed through adaptation was found to enhance the grain yields of clusterbean by 10 to 13 % even under low to moderate drought conditions. In a field experiment adoption of an organic mulching of weeds at 6 t ha^{-1} in between rows during a drought year led to a significant improvement in nodulation and N_2-ase activity of arid legumes resulting in an enhanced grain production and straw yields. This effect might be due to the availability of more moisture and creation of congenial physical conditions near the root zone compared to that to the unmulched fields.

Microbial mobilization of soil phosphorus

Arid soils contain a high amount of phosphorus, i.e., 557-729 kg ha^{-1} of which only 2.4 –3.9 % is present in available form depending upon the land use pattern while 15-20% (97-110 kg ha^{-1}) of the total P is present as organic form such as phytin, lecithin, phospholipids and other unidentified compounds (Tarafdar *et al.*, 1989). But the response of legumes including clusterbean to P fertilization in arid soil is variable (Kathju *et al.,* 1987). In order to mobilize soil organic P, some soil fungi producing significantly higher amounts of phosphates were identified and referred to as phosphatase producing fungi (PPF). Upon inoculation, the PPF were found to enhance the dry matter production and grain yield of clusterbean (Tarafdar *et al.*, 1995) through the improvement in the activities of various enzymes in the rhizosphere and better nutrient availability besides the production of growth regulators by PPF. Further there was a significant improvement in the uptake of

N, P, K, Ca, Mg, Fe and Zn while there was no effect in the uptake of Na, Mn and Cu.

Arbuscular mycorrhizal fungi are ubiquitous and are found in association with all crop plants including clusterbean (Kiran Bala *et al.*, 1989) reported that the percent root infection by AM fungi in clusterbean varied from 30-60%. The infection rate was increasing from the first week onwards with the maximum infection at 5 weeks after sowing and the started declining. Upon soil inoculation with different AM-fungi there was a significant improvement in the dry weights of shoots and roots (Champawat and Somani, 1990; Rao and Tarafdar, 1993, Tarafdar and Rao, 1997). AM fungi under field conditions in micro-plots significantly enhanced nodulation and nitrogenase activity of clusterbean (Table 5). This might be due to better P-nutrition through tapping a large volume of soil beyond P-depletion zone by the ramifications of AM-fungal hyphae or to the increased availability of P in the rhizosphere soil resulting from the break-down of organic phosphates by the phosphatases secreted by AM- fungi or AMF infected plants. The enhancement in per cent root infection and build up of viable AMF spores in the rhizosphere soil was also observed. AM-fungal inoculation had significantly improved the microbial activity in the rhizosphere. All the effects led to the significant improvement in the dry matter production and grain yield of clusterbean upon inoculation with AM-fungi. This could be attributed to the extensive root system tapping a larger volume of soil for water and nutrients. Uptake of various nutrients especially N and P was improved significantly upon AM-fungal inoculation.

Table 5 Effect of AM-fungi on clusterbean

Parameter	Control	Inoculated with	
		GM	GF
Nodule No. ($plant^{-1}$)	16.5	21.0	20.8
N_2-ase activity*	385	510	490
Root infection (%)	35	83	78
AMF spores($100g^{-1}$)	352	916	725
Grain yield (kg ha^{-1})	540	700	620
Straw yield (kg ha^{-1})	2300	3170	2990

Source: Tarafdar and Rao, 1997

* η moles C_2H_4 $Plant^{-1}h^{-1}$, GM: *Glomus mosseae*, GF: *G. fasciculatum*

Dual inoculation of *Rhizobium* and AM-fungi with FYM application

Inoculation of clusterbean with *Rhizobium* and arbuscular mycorrhizal fungus coupled with the application of FYM at 4 t ha^{-1} under field conditions significantly improved nodule number, nitrogenase activity, per cent root infection and density of viable AM-fungal spores. Similarly, grain yield and dry matter production of clusterbean were improved upon inoculation consistently for two years with the maximum enhancement with dual inoculation coupled with FYM application. The effect of seed inoculation with *Rhizobium* and soil inoculation with AM-fungus was equivalent to that of 20 kg N and 40 kg P_2O_5 ha^{-1}, respectively. Application of FYM had additive effect on dry matter production and grain yield of clusterbean. In the light of the above it is suggested that biological management of arid soils through introduction of efficient strains of symbiotic nitrogen fixing bacteria and AM-fungi coupled with the application of FYM is of great significance in obtaining sustainable clusterbean production and improving the biological productivity of the arid soils.

Significance of clusterbean in sustainable crop production

Sustainable plant production can be achieved through the adoption of such technologies that enhance the soil biological productivity. It is the soil productivity developed / maintained by enhancing the natural processes through biological means. It provides balanced nutrition through steady turnover of organic matter, release of nutrients in harmony with the needs of the plants and non-availability of nutrients in excessive amounts. It was observed that monitoring the status of biological productivity and the functioning of soil ecology may provide useful information in assessing the sustainability of an agricultural system. The status of biological productivity can be measured by determining various components such as soil organic matter, activities of various enzymes (dehydrogenase and phosphatases), nitrogen fixing potential of soil (N_2-ase activity), populations of nitrifying bacteria and viable AM-fungal spore density. Soil biotechnological approaches for achieving sustainable agricultural production through the enhancement in the soil biological productivity are use of bioinoculants, use of the organic manures, adoption of INM practices and adoption of various farming systems (crop rotation, inter/ mixed cropping, agro-forestry, silvi-pasture, agro-horticulture, ley farming, alley cropping etc).

Table 6. Pearl millet production after 3 years of crop rotation in an arid soil

Rotation	Yield (kg ha^{-1})	
	Grain	Straw
F-PM-F	340	1680
PM-PM-PM	380	1760
MB-PM-MB	440	1900
CB-PM-CB	530	1980
F-MB-MB	560	1940
F-CB-CB	530	1940
MB-MB-MB	630	2200
CB-CB-CB	680	2320
LSD (P=0.05)	80	310

Source: Rao *et al.,* 1995
Fallow; PM: Pearl millet; MB: Mung bean, CB: Clusterbean

These cropping systems support the recycling of nutrients particularly nitrogen and maintain the proper balance between organic matter and biological activity which are the important components of a productive soil. Inclusion of a legume, clusterbean in a pearl millet based crop rotation for one, two or three years had improved various parameters of biological productivity of soil (activities of various enzymes, organic matter, nitrifying bacteria, AM-fungal spores etc.) over a continunous fallow or continuous pearl millet resulting in a significant increase in pearl millet production. (Table 6). Continuous clusterbean for 3 years in rotation gave the maximum increase in soil biological productivity resulting in significantly higher production of pearl millet followed by 2 or 1 year legume and continuous pearl millet.

Future strategy

In view of the above the benefits on the use of beneficial microorganisms for sustainable clusterbean production have not been harvested to its full potential because of lack of temperature and desiccation tolerant strains. As clusterbean is grown in soils containing high amounts of unavailable phosphorus there is a need to develop strains of micro-organisms (AM fungi, P-solubilizers etc) that can mobilize soil phosphorus for higher production in clusterbean. In conjunction with such strains of micro-organisms suitable agronomic practices and cropping systems that may enhance the population of beneficial micro-organisms need to be identified. Some attempts are needed to develop genetically engineered host and micro-organisms that can tolerate high temperature and desiccation for sustainable clusterbean production in the harsh environment. There is a need to extensively study the role of organic manures along with the use of biofertilizer (*Rhizobium*, phosphorus mobilisers) for achieving sustainable clusterbean production and at the same to maintain the soil health. There is a need to study extensively the role of organic manures coupled with biofertilizers for achieving clusterbean production and at the same time improving the soil health.

SUMMARY

Nodulation and nitrogen fixation in clusterbean under natural conditions are very poor and sometimes absent due to lack of optimum soil temperature and moisture and population of suitable micro-symbiont in these soils. Some efficient strains of *Rhizobium* have been isolated from arid soils and found to improve nodulation and nitrogen fixation upon seed inoculation resulting in higher dry matter production and grain yields even under moderate stress levels. With imposition of water stress, the reduction in N_2-ase activity was lowest in clusterbean compared to other legumes. Water stress during vegetative stage may be more detrimental for nitrogen fixation. Nitrogen fixation in clusterbean is highly sensitive to soil moisture stress but it has a higher capacity to recover under favourable moisture conditions. Moisture stress effects may be alleviated by using stress tolerant strains of *Rhizobium*, improving the method of inoculation and pelleting with synthetic polymers and adoption of soil moisture conservation techniques such as organic mulching. Further Phosphorus requirements of this crop can be met through the use of phosphatase producing fungi or arbuscular mycorrhizal fungi. Sustainable clusterbean production can be achieved with dual inoculation of efficient strains of *Rhizobium* and AM-fungi coupled with the application of FYM. Continuous clusterbean compared to any other legume for three years provide maximum enhancement in soil biological productivity resulting in significantly higher production of pearl millet.

REFERENCES

Ahlawat, O.P., Sharma, H. R. and Dadarwal, K.R. 1992. Bacteriocin producing mutants of cow pea miscellany *Rhizobium* role in strain competition. Indian J. Microbiol. 32 : 263-268.

Akhavan-Kharazian, M., Campbell, W.F. Jurinak, J.J. and Dudley, L.M. Calcium amelioration of NaCl effects on plant growth, chlorophyll and ion concentration in *Phaseolus vulgaris*. Arid Soil Research and Rehabilit., 5: 9-19.

Anonymous. 1980. Annual Progress Report, CAZRI, pp. 130-156.

Arayangkool, T., Schomberg, H.H. and Weaver, R.W. 1990. Nodulation and N_2-fixation of guar at high root temperature. Plant and Soil, 126: 209-213.

Balasubramanian, V. and Sinha, S.K. 1976. Effect of salt stress on growth, nodulation and nitrogen fixation in cow pea and mung bean. Physiol. Plant., 36: 197-200.

Beadle, N.C.W. and Tchan, Y.T. 1953. Nitrogen economy in semi-arid plant communities, Part I. The environment and general considerations. Proc. Linn. Soc. NSW., 80: 62-70.

Bernstein, L. and Ogata, G. 1966. Effect of salinity on nodulation, nitrogen fixation and growth of soybean and alfalfa. *Agron. J.*, 58: 201-203.

Bell, R.W., Edward, B.G. and Asher, C.J. 1989. External calcium requirements for growth and nodulation of six tropical food legumes grown in flowing culture solution. *Aust. J. Agri. Res.*, 40 : 85-96.

Bhardwaj, K.K.R. 1975. Survival and symbiotic characteristics of *Rhizobium* in saline alkali soils. Plant and soil, 43: 377-385.

Bhupinder Singh, Chatterjee, S.R, Nair T.V.R. and Singh, B. 1994. Effect of light and darkness on root nodule metabolism in clusterbean (*Cyamopsis tetragonologa* (L.) Taub.) Journal Nuclear Agriculture and Biology. 23: 224-229.

Brockwell, J. 1976. Ecology of legume- *Rhizobium* associations. Proc Ind Nat Sci. Aca. 40 B, 687-699.

Champawat, R.S. and Somani, L.L. 1990. Effect of vesicular-arbuscular mycorrhiza and phosphorus fertilizer on cluster bean. Trans. Ind. Soc. Desert Technol., 15: 17-22.

Dadarwal, K.R, Sindhu, S.S., Batra, R. and Renu Batra. 1985. Ecology of Hup^+ *Rhizobium* strains of cow pea miscellany: native frequency and competence. Arch Microbiol., 141: 255-259.

Dart. P. J., Day, J., Islam, R. and Dobereiner, J. 1976. Symbiosis in tropical grain legumes: some effects of temperature and the composition of the rooting medium. In : *Symbiotic Nitrogen Fixation in Plants* (ed. Nutman PS) pp 361-384. IBP No. 7 Camb. Univ. Press. Cambridge.

Elsheikh, E.A.E. and Wood, M. 1990. Salt effects on survival and multiplication of chick pea and soybean rhizobia. Soil Biol. Biochem., 22: 343-347.

Engin, M and Sprent, J.I. 1973. Effect of water stress on growth and nitrogen fixing activity of *Trifolium repens*. New Phytol. 72: 117-126.

Erdman, L.M. 1948. Strains of *Rhizobium* effective on guar. J. Amer. Soc. Agron., 40: 364-369.

Francois, L.E., Donovan, T.J. and Mass, E.V. 1990. Salinity effects on emergence, vegetative growth and seed yield of guar. Agron. J., 82: 587-591.

Garg, B.K. Venkateswarlu, B., Vyas, S.P. and Lahiri, A.N. 1984. Nodulation and N_2-ase activity in clusterbean. Proc. Indian Acad. Sci. (Pl. Sci.), 96: 531-538.

Garg, B.K., Kathju, S., Vyas, S.P. and Lahiri, A.N. 1986. Effect of saline waters on drought affected clusterbean. *Proc. Indian Acad. Sci. (Pl. Sci.)* 96 : 531-538.

Garg, B.K., Kathju, S., Vyas, S.P. and Lahiri, A.N. 1997. Alleviation of sodium chloride induced inhibition of growth and nitrogen metabolism of clusterbean by calcium. Biol. Plant., 39: 395-401.

Garg. R.P. and Garg. F.C. 1993. *Ex-planta* nitrogenase activity of *Bradyrhizobium* spp. (pigeonpea, clusterbean). Ann. Biol. 9 : 188-193.

Garg, R.P., Kukreja. and Garg, F.C. 1993. Efficiency of a cross infective *Rhizobium* strain in heterologous hosts. Environ. Ecol., 11: 256-259.

Ghonsikar, C.P. and Saxena, S.N. 1993. Influence of molybdenum and tryptophane on nodulation in clusterbean. (*Cyamopsis tetragonoloba*) Indian J. Agric. Sci., 938-941.

Girija, V.K. and Nair, S.K. 1985. Occurrence of vesicular-arbuscular mycorrhiza in certain crop plants of Kerala. Agrl. Res. J. Kerala, 23: 185-188.

Huang, C.Y., Boyer, J.S. and Vandernoef, L.N. 1975. Limitation of acetylene reduction (nitrogen fixation) by phosphate in soybean having low water potentials. Plant Physiol. 56: 228-332.

Kaur, A., Sheoran, I.S. and Randhir Singh. 1985. Effect of water stress on the enzymes of nitrogen metabolism in mung bean (*Vigna radiata* (L.) Wilczek) nodules. Plant Cell and Environ., 8: 195-200.

Kathju, S. Aggarwal, R.K. and Lahiri, A.N. 1981. Evaluation of diverse effects of phosphate application on legumes of arid areas. Trop Agric., 64: 91-96.

Khurana, A.L., Saini, M.L., Jhorar, B.S. and Dudeja, S.S. 1978. Nodulation studies in clusterbean (*Cyamopsis tetragonoloba* (L.) Taub). Forage Res., 4: 199-202.

Kiran Bala, Rao, A.V. and Tarafdar, J.C.. Occurrence of VAM associations in different plant species of the Indian desert. Arid Soil Res. Rehabil., 3: 391-396.

Kumar, A. 1987. Note on nodulation and nitrogen fixation efficiency of guar under salinity. Curr. Agri., 11: 85-86.

Kuo, T. and Boersma, L.L. 1971. Soil water stress and root temperature effect on nitrogen fixation in soybean. Agron. J., 63: 901-904.

Lakshmikumari, M., Singh, C.S. and Subba Rao, N.S. 1974. Root hair infection and nodulation in lucehne (*Medicago sativa*) as influenced by salinity and alkalinity. Plant Soil, 40: 261-268.

Lahiri, A.N., Garg, B.K., Kathju, S., Vyas, S.P. and Mali, P.C. 1987. Responses of clusterbean to soil salinity. Ann. Arid Zone, 26: 33-42.

Lynd, J.Q. and Ansman, T.R. 1989. Soil fertility effects on nodulation, nitrogen fixation and yields of guar *(Cyamopsistetragonoloba* (L.) Taub.) grown on a typic Eutrustox. J. Pl. Nutri., 8: 1069-1084.

Mand, S., Dahiya, B.N. and Lakshminarayana, K. 1991. Nodulation, nitrogen fixation and biomass yield by slow and fast growing cow pea rhizobia in guar under different environments. Ann. Biol., 7: 31-37.

Nandwal, A.S., Dabas, S., Bharati, S. and Yadav, B.D. 1990. Zinc effect on nitrogen fixation and clusterbean yield. Ann. Arid Zone, 29: 99-103.

Oke, O.L. 1967. Nitrogen fixing capacity of guar beans. Trop. Sci. 9 :144-147.

Raju, M.S. 1936. Studies on the bacterial-plant groups of cow pea. *Cicer* and Dhaincha. Zentl Bakt., 94: 249-262.

Rao, A.V. 1995. Biological nitrogen fixation in arid ecosystem. *In: Plant- Microbe interactions in sustainable agriculture.* (eds Behl, R.K. et al.) pp 89-101. Bioscience publishers, Hisar.

Rao, A.V. and Tarafdar, J.C. 1993. Role of VAM-fungi in nutrient uptake and growth of clusterbean in an arid soil. Arid Soil Rehabili., 7: 275-280.

Rao, A.V. and Venkateswarlu, B. 1983. Pattern of nodulation and nitrogen fixation in moth bean. Indian J. Agric. Sci. 53: 1035-1038.

Rao, A.V., Venkateswarlu, B. and Henry, A. 1984. Genetic variation in nodulation and nitrogenase activity in guar and moth. Indian J. Genet. Pl. Breed., 44: 54-57.

Rao, A.V. and Venkateswarlu, B. 1987. Nitrogen fixation as influenced by water stress in selected crop legumes of the Indian Arid Zone. Arid Soi Res. Rehabilit., 1 : 89 - 96.

Rao, A.V., Tarafdar, J.C., Sharma, S.K., Praveen Kumar and Aggarawal, R.K., 1995. Influence of cropping systems on soil biochemical properties in an arid rainfed environment. J. Arid Environ., 31 : 237-244

Raut, M.S. and Ali, M. 1983. Response of clusterbean (*Cyamopsis tetragonoloba*) to phosphate application and rhizobial inoculation under dryland conditions. Legume Res., 6: 65-68.

Richmond, T.E. 1926. The nodule organism of the cow pea group. J. Am Soc. Agron., 18: 411-414.

Sprent, J.I. 1971. Effects of water stress on nitrogen fixation in root nodules. Plant Soil Special. Volume. 225-228.

Sprent, J.I. 1973. Growth and nitrogen fixation in *Lupinus arboreus* is affected by shading and water supply. *New Phytologist*, 72 : 1005-1022.

Sprent, J.I. 1976. Water stress and nitrogen fixing root nodules. *In: Water deficits and plant growth* (ed. T.T. Kozlowski) Vol IV pp. 291-315. Academic Press, New York.

Sprent, J.I. 1984. Effects of drought and salinity on heterotrophic nitrogen fixing bacteria and on infection of legumes by rhizobia. In: *Advances in Nitrogen Fixation Research* (eds. C. Vajeer and W.E. Newton) pp. 295-302. Martinus Nijhoff. The Hague.

Stafford, R.E.and Lewis, C.R. 1980. Nodulation in inoculated and non-inoculated Kinman guar. Prog. Rep. Texas Agric. Exp. Stat. 165-170.

Swaraj, K., Dabas, S. and Sheoran, I.S. 1988. Diurnal variation in symbiotic nitrogen fixation (acetylene reduction activity) in some legumes. Ann. Biol., 4: 109-111.

Tarafdar. J.C. and Rao, A.V. 1997. Response of arid legumes to VAM-fungal inoculation. Symbiosis, 22: 265-274.

Tarafdar, J.C., Kiran Bala and Rao, A.V. 1989. Phosphatase activity and distribution of phosphorus in arid soil profiles under different land use patterns. J. Arid. Environ. 16: 29-34.

Tarafdar, J.C., Rao, A.V. and Praveen Kumar. 1995. Role of phosphatase producing fungi on growth and nutrition of clusterbean (*Cyamposis tetragonoloba* (L.) Taub) J. Arid Environ., 29: 331-337.

Tu, J.v. 1981. Effect of salinity on *Rhizobium* –root hair interaction, nodulation and growth of soybean. Can. J. Pl. Sci., 61: 231-239.

Venkateswarlu, B. and Rao, A.V. 1987. Quantitative effects of field water stress on N_2 (C_2H_2) fixation in selected legumes grown in Indian desert. *Biol. Fert. Soil.* 5: 18-22.

Venkateswarlu, B. Rao, A.V. and Lahiri, A.N. 1982. Symbiotic performance of rhizobial strains on guar (*Cyamopsis tetragonoloba* (L.) Taub.) grown in a desertic soil. *Egypt J. Microbiol.,* 18: 9-14.

Venkateswarlu, B., Rao, A.V. and Lahiri, A.N. 1983. Effect of water stress on nodulation and nitrogenase activity of guar (*Cyamopsis tetragonoloba* (L.) Taub.) *Proc. Indian. Acad. Sci. (Plant Sci.),* 92: 297-301.

Vyas, S.P., Kathju, S., Garg, B.K. and Lahiri, A.N. 1996. Response of clusterbean genotypes to shade. *Indian. J. Plant Physiol.* (New Series), 1: 2324-238.

Wani, A.G. and Patil, B.B. 1982. Response of clusterbean (*Cyamopsis tetragonoloba* (L.) Taub.) varieties to phosphate fertilization and bacterial inoculation. J *Maarah Agric. Univ.,* 7: 63-65.

West, N.E. 1975. Short term Ecosystem Dyanamics in Arid Lands. Camb Univ. Press, London, pp. 272.

Yadav, N.K. and Vyas, S.R. 1971. Response of root nodule rhizobia to saline alkaline and arid conditions. *Indian J. Agric. Sci.,* 41: 875-881.

CHAPTER - 8

DISEASES

Satish Lodha and Ritu Mawar

INTRODUCTION

The productivity of guar has been very low and sometimes becomes uneconomic probably due to erratic and uncertain rains, and cultivation of traditional strains, which are late maturing, poor yielder and highly susceptible to a number of diseases. To increase the guar yields to a maximum and to improve the quality, it is imperative that we increase our knowledge of many diseases, which affect the quality and yield of the crop. In the present review efforts have been made to compile available information on diseases of this crop.

FUNGAL DISEASES

1. Alternaria leaf spot

This disease was reported in Pusa (Bihar) and Madras. It is caused by *Alternaria brassica* (Berk.) Sacc. (Butler, 1918; Streets, 1948), by *Alternaria cyamopsidis* Rangaswami and Rao (Rangaswami and Rao, 1957) and by *A. Cyamopsidis* (Ell. And Ev.) Elliot (Sowell, 1965a). Orellana and Simmons (1966) proposed that *A. cyamopsidis* be systematically treated as a variety of *A. cucumerina* and suggested a new combination, *A. cucumerina var. Cyamopsidis*. The fungus attacks clusterbean wherever the crop is grown in India, Pakistan, Africa and in the United States (Butler, 1918; Orellana and Simmons, 1966;

Westhuizen and Beck, 1987). It was reported to be seed borne and may have been introduced into the United States on seeds from India (Sowell, 1965a).

The disease manifests itself primarily on leaf blade as dark brown, round to irregular spots varying from 2 to 10 mm in diameter. In the early stages of infection, water soaked spots appear on leaf blade which later on turn grayish to dark brown with concentric zonations, demarked with light brown lines. On the under surface, the spots are light to grayish brown. In severe infection, several spots merge together involving a major part of the leaf blade. In such cases, the leaflets become chlorotic and usually drop off. If the plants are infected in the early stages of growth, they may not flower, thereby causing severe losses. High yield losses (43-87%) were recorded when leaves were infected at seedling stage than at old age (Sharma, 1983).

The *in vitro* growth requirements of *A. cyamopsidis* were studied by Singh and Prasada (1973). Potato Dextrose Agar and Richard's medium supported maximum growth while maximum sporulation was observed on corn meal Agar. Among carbon and nitrogen sources, trace elements and vitamins tested, fructose, phenylalanine, iron and thiamine supported maximum growth. A temperature of 30°C, pH 6.0 and 12 h cycles of light and darkness supported maximum growth (Ayub *et al.,* 1995). Excessive rainfall and humidity favour the incidence of Alternaria blight during the growing season. The maximum severity of the blight takes place between bloom and pod set.

Alternaria may be controlled by sowing pathogen free seeds or cultivars resistant to the pathogen. Chand and Verma (1968b) reported two varieties CS 114 and CS 115 resistant to *A. tenuissima*. Gandhi *et al.,* (1978b) also reported resistance in some genotypes. Early sown crop contracted maximum infection. Among fungicides, zineb (0.2%) and copper oxychloride proved most effective as two sprays at an interval of 15 days in controlling disease (Mathur *et al.,* 1972). A combination of zineb with streptocycline has been recommended for simultaneously controlling bacterial blight and Alternaria blight.

2. Powdery Mildew

This disease was first reported from Bombay and Madras in India (Butler, 1918) and Pakistan (Khan, 1958). The associated pathogen is also known to cause powdery mildew on many other plant species. It is widely distributed in South and Central Asia, Northern Africa and South and Central Europe. Other than *O. taurica, Erysiphe polygoni* (Ratnam *et al,*, 1985) and *Sphaerotheca fuliginea* (Chandra and Saxena, 1990) are also reported from India to cause powdery mildew.

The disease manifests on the leaves only. The affected leaves are covered with mycelial patches dotted with the fruiting bodies (perithecia). Severely affected plants are defoliated and weakened by premature drying and death of infected leaves. It has been observed that the incidence of powdery mildew is more prevalent in the regions where crop season is prolonged. Warm temperatures (33°C and above), low humidity (below 75%) and bright sunshine are congenial for the disease development. Thus in India, it is more common in Karnataka, Andhra Pradesh and Gujarat compared to Rajasthan and Haryana. The perithecial stage of the fungus first appears on the dried leaves. Venkatarayan (1946) reported that *Cicinnobolus cesatii* De Bary is a natural fungus parasite of powdery mildew. It has been reported that under favourable conditions the mildew causes considerable defoliation, so that the inhibitory properties of its parasites are evident only moderate, though probably sufficient to arrest the full development and spread of the fungus.

Reddy and Rao (1971) reported effective control of powdery mildew with Benlate (0.025%). Sulphur compounds, cosan, thiovit and sultaf have inhibited germination of the conidia of *O. taurica* in *in vitro* studies (Mathur and Batrice, 1971). Satisfactory disease control have also been achieved with two sprays of N-triaecyl-6, 6-dimethyl morpholine (0.05%) at 15 days interval (Solanki and Singh, 1976). In other studies, Tridemorph sprays controlled powdery mildew caused by *O. taurica* and increased the green pod yield (Sharma, 1984), while Bavistin was found effective against *E. polygonii* (Ratnam et al., 1985). Studies conducted under AICRP on arid legumes at Gujarat indicated that seed treatment with Streptocycline (0.025%) and spraying a combination of streptocycline and Dinocap (both 0.01%) was effective in controlling powdery mildew as well as bacterial blight (Table 1). No adverse effect was noticed in combining either Dinocap or Triadimefon with streptocycline. Genotype GAUG 63 possessed some what resistance to powdery mildew.

3. Dry Root rot

This disease is most serious under dry and warm growing

conditions. Prasad (1945) reported that *Macrophomina phaseolina* also caused root rot in guar, besides in cotton. The disease may occur at any stage of the crop from pre-emergence to maturity and can appear as seedling blight, ashy stem blight or dry root rot. Seedling blight occurs primarily on the cotyledons in the form of damping off or as elongated block cankers on growing seedlings. Infected seedlings can be easily uprooted. In grown up plants, the disease appears as bronzing of leaves on one or more branching followed by drooping of the upper tender parts of the shoot. Affected plants continue to grow without having clusters or pods, but may break easily in strong winds. Nodule formation in these plants is also drastically reduced. Heavily infected plants rarely produce any grain and if produced seeds are also infected.

Studies have shown that highest population of *M. phaseolina* (233 sclerotia g-1 soil) after the harvest of clusterbean crop was at 0-5 cm depth followed by that at 20-30 cm (150 g^{-1} soil), 10-20 cm (93 g^{-1} soil) and 5-10 cm layer (63 g^{-1} soil). Presence of relatively high population in the surface soil was attributed to high soil temperature (35-40°C) and low soil moisture content (0.5 to 1%). Rotation with less susceptible crops like, moth bean and pearl millet or keeping the land fallow reduced the population of *M. phaseolina* (Lodha et al., 1990). Another approach is to go for *M. phaseolina* resistant genotype viz., Kutch-8 and RGC 471 (Lodha and Solanki, 1993).

Efficiency of moisture conservation techniques in reducing soil borne inoculum and disease intensity was also ascertained. Mulching of the soil with pearl millet stover (3.5 ton ha^{-1}), low plant density (1.6 against 2 lakh ha^{-1}) and farmyard manure (10 ton ha^{-1}), singly or in combination effectively conserved the soil moisture and reduced the population densities of *M. phaseolina* and dry root rot incidence (Lodha, 1996). Water relation parameters also play pivotal role in governing tolerance of genotypes to *M. phaseolina* infection. Among arid legumes, cowpea plants became predisposed to fungal infection only after its shoot water potential declined below – 1.4 MPa, however clusterbean and moth bean reached that stage at –1.8 and 2.0 MPa, respectively. Incorporation of zinc or copper (5 kg ha^{-1}) or sulphur (10 kg ha^{-1}) coupled with farmyard manure also brought down dry root rot incidence and increased the yield of clusterbean (Akem and Lodha, 1997).

Efficacy of soil solarization was also ascertained against *M. phaseolina*. Mulching the soil with transparent polyethylene sheet (50 g) during May-June considerably increased the soil temperature and killed the viable propagules of *M. phaseolina* within 15 days. Clusterbean crop raised in solarized plots remained disease free till harvest. Significant reduction in weed population and increase in available contents of potassium and calcium were also recorded in solarized plots, resulting in considerable increase in seed yield of clusterbean (Lodha, 1995).

In summer months, merely by the natural heating of the bare soil led to almost 42% reduction in the population of *M. phaseolina.* Incorporation of N plus manure at the time of natural heating further reduced the population of *M. phaseolina* (Lodha 1995). Substituting these amendments by cruciferous residues like mustard cake (0.18%) and cauliflower (0.20%) further augmented reduction in the population by about 74-76% within 30 cm soil depth (Lodha et al., 1997). Increased antagonism against *M. phaseolina*, particularly by actinomycetes in amended soils also induced soil suppressiveness (Sharma *et al.*, 1995).

Amendment of compost prepared from pearl millet and cauliflower leaf residues (4t ha^{-1}) increased nutrients and microbial population in the soil that decreased the population of *M. phaseolina* (Praveen Kumar *et al.*, 1998). As a result, minimum root rot mortality and maximum seed yield was also recorded in amended plots. A schematic diagram of various technologies for management of dry root rot are illustrated in Fig. 1.

4. Phymatotrichum Root Rot

The root rot caused by a soil-borne fungus, *Phymatotrichum omnivorum* has been reported to be prevalent in heavy, alkaline soils in parts of south-western United States. However, in Texas, guar was reported resistant to *P. omnivorum*, but it was found that most infected plants in Texas survived upto maturity and that the infection could be detected only by pulling the plants from the soil. In Arizona, 10-18% losses were observed due to this disease. The wilting and death of plants in the later part of the growing season may be due to infection of fungus resulting in death of the roots. The disease has not been reported from any other part of the world.

5. Leaf Spots

(a) Cercospora Leaf Spot – Leaf spot caused by *Cercospora psoraleae* was first time reported from India

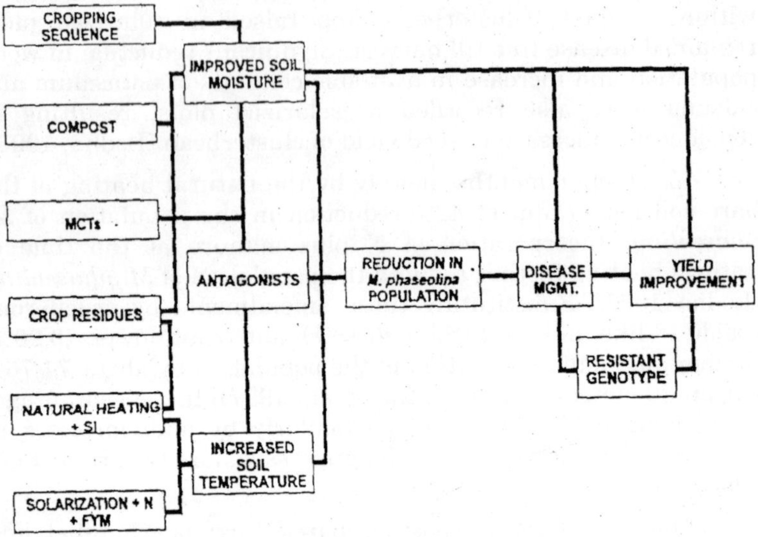

Fig. 1. Integration of Technologies for Management of Dry Root Rot and Yield Improvement.

(Vasudeva, 1960). It was observed that some seeds develop a purple discoloration and dark effuse lesions occur on the stem and leaves in many cases. These symptoms resemble those occurring on Soybean. Isolates from soybean and guar are cross pathogenic on inoculation and cause similar symptoms. The guar isolate also cause leaf spotting of *Psoralea sp.* but differ morphologically from *C. psoraleae*.

(b) Myrothecium Leaf Spot – The leaf spot caused by *Myrothecium roridum* was reported first time from Jodhpur, Rajasthan (Arya, 1956). The disease appears as minute oil soaked spores 12 mm in diameter which become brown in color. Under favourable environmental conditions like temperature and humidity, these spots become as large as 10-15 mm and give characteristic target board type of appearance with distinct circular zonations. Often these spots may coalese and cover a fairly large area of the leaf. In severe cases of infection the entire leaf becomes chlorotic and finally gets defoliated. The disease has also been reported to cause pod spot (Shivanna and Shetty, 1986). Seeds after infection loose their viability. The plants with turgid leaves are found comparatively more susceptible. Isolates of *M. roridum* from guar and seven other hosts near Jodhpur varied in pathogenicity but were not host specific (Arya, 1959).

(c) Curvularia Leaf Spot – A leaf spot caused by *Curvularia lunata* was first reported from Hisar, India (Chand and Verma, 1968a). The infection starts in the form of small circular to oval, scattered spots on the leaf lamina, occasionally these spots may coalese occupying large areas. Finally, the leaves show a typical blighted appearance and drop down. No other part of the plant shows any symptom. Singh et al., (1974) found maximum disease control when Aureofungin was sprayed twice at an interval of 15 days. Genotype EC 248/P3 appeared tolerant against this disease.

(d) Anthracnose – A disease characterized by black spots on leaves petiole and stem caused by *Collelotrichum capsici f. cyamopsicola* was observed on the glabrous type guar at Anand, India during rainy seasons of 1952 and 1953 (Desai and Prasad 1955). A severe blight due to this pathogen has also been reported from Rajasthan (Kothari and Bhatnagar, 1966) and Andhra Pradesh (Rao and Rao, 1956). Long black streaks on stems are the most conspicuous symptoms. In later stages, splitting of the stem parallel to the axis and along the ridge take place. Severely infected plants remain sunted in growth and dry up. Hairy varieties generally show less disease intensity.

Best *in vitro* growth of *Colletotrichum* isolates was obtained with 0.5 ppm of maleic hydrazide and 1-naphthyl acetic acid, 0.05 ppm of 2-naphtoxy acetic acid and 1 ppm of indol-3-yl-acetic acid (Agnihotri and Prasad, 1966). Results of *in vitro* tests of Kothari and Bhatnagar (1966) showed that Ferbam inhibited spore germination completely at 2 ppm level.

6. Blights

(a) Ascochyta Leaf Blight – *Ascochyta imperfecta* causing black stem disease of alfa-alfa was also found pathogenic on guar (Kernhamp and Hemerick, 1953). In earlier stage of infection, small black lesions appear on leaves and stem. With the advancement of the disease profuse defoliation occurs while stem lesions enlarge and coaleses until the entire stem becomes surrounded by a black sheet of fungus mycelium and diseased host tissues. Peduncles and floral parts are generally infected and flowers frequently fall down.

(b) Fusarium Blight – The blight caused by *Fusarium moniliforme* (Sheild) Snyder and Hansen, was first reported from Anand (Prasad and Desai, 1952). It usually appears in late sown crop and the symptoms are more conspicuous on hairy varieties. The disease starts as black streaks on stem, which extends both upwards and downwards. The pedicels of the leaves are also affected and gradually the shoot. In severe cases the entire plant presents a wilted appearance. Intercellular fungal hyphae are seen within the parenchymatous tissues of the infected part. Incidence of the disease is less in well drained sandy soils compared to ill drained black soils.

(c) Rhizoctonia blight – Blighting and rotting of leaves, petioles and young branches of guar were observed due to *Rhizoctonia solani* Kuhn. The disease is destructive in the month of August during drizzling weather coupled with high temperature. It appears on freshly emerged seedlings under waterlogged conditions as small, with circular to irregular, pale water soaked spots on thick cotyledons.

BACTERIAL DISEASES

1. Bacterial blight

This is the most serious disease of clusterbean and was first recorded in India from Patna (Patel *et al.*, 1953) and named as *Xanthomonas cyamopsidis*. Later on the same diseases was identified as bacterial blight (Patel and Patel, 1958; Srivastava and Rao, 1963a). Since then, this disease has also been reported from U.S.A. (Orellana *et al.*, 1965).

The disease is widespread in all the clusterbean growing regions and occurs in severe form particularly in monsoon season crop (Chand and Gandhi, 1978). The extent of damage depends on stage of the crop, genotype and the prevailing environmental conditions. Gupta (1978) reported 58% reduction in yield under artificial inoculation conditions. Lodha and Gupta (1981) recorded yield loss of 47-50% at 69-72% blight intensity under arid conditions and observed that number of pods were most severely affected with increasing disease intensity. Gandhi and Chand (1985) noticed that under semi-arid conditions, clusters/plant were also adversely affected due to blight.

The disease appears as both leaf spots and blight simultaneously. The spots resulting from stomatal infection are intraveinal, round, water soaked or oily in appearance and well defined on the dorsal surface of the leaf. The spots may enlarge or coalesce but ultimately result in blight phase. Hydathodal infections give rise to V-shaped lesions starting from margins and progress towards the midrib. From the leaf, infection advances systematically through petioles into the stem and produces longitudinal streaks that result into blackening and cracking of the stem. Black streaks sometimes may appear on pods also due to local infection.

The causal agent, *Xanthomonas campestris pv. Cyamopsidis* is rod shaped, borne singly or in short chains 0.8, 1.7 um in size, Gram negative, capsulated, motile with a single polar flagellum and non-spore forming. The organism grows well between 27-35°C. Two races of the pathogen race '0' and race '1' have been reported from USA (Orellana and Weber, 1971). On the basis of soluble protein pattern, existence of two races have been indicated from Haryana, India (Parashar *et al.*, 1994). The pathogen could also infect french bean, while under artificial inoculation conditions *Aseratum conyzoides*, a weed also got infection by pathogen. The pathogen enters the leaves through stomata, hydathodes or wounds and is largely vascular. The occurrence of disease is favoured by spattering rains, high humidity (80-90%) and moderate temperature (27-30°C) which are usually available during monsoon months (Aug.-Sep) in India (Singh and Swarup, 1987). The crop stage between 4-6 weeks is most susceptible and there is a negative correlation between disease severity and host stage (Singh and Swarup, 1987).

The bacterium is internally seed borne (Srivastava and Rao, 1963b). Seedlings from diseased seed develops cotyledonary lesions soon after emergence. Survival of bacterium in infected plant debris or in soil could not be observed (Singh *et al.*, 1988). Jain and Chakravarti (1988a) evaluated different methods of detecting *Xanthomonas* from clusterbean seeds and reported that growing infected seeds in quartz sand gave maximum infection compared to agar plate and test tube agar method.

Mali *et al.* (1989) analysed the biochemical components of resistant and susceptible clusterbean genotypes and observed that genotypes differed in metabolic constituents upon infection. With severe infection amount of reducing sugar increased in all the genotypes studied. However, the per cent increase was highest in resistant genotype (HG75) and lowest in highly susceptible genotype (Pusa Navbahar). They also suggested the possibility of using glutamine synthatase as an additional parameter for determining relative resistance in genotypes. Lodha *et al.* (1993) observed maximum blight intensity (76%) and apparent infection rate (0.052) in susceptible genotype 'Pusa Nav bahar'. Lesser change in soluble carbohydrate, protein and chlorophyll coupled with more changes in phenols and peroxidase due to blight infection in resistant, genotype 'HG75' compared to 'Pusa Nav bahar' partly accounted for the bacterial blight resistance in 'HG75'. Higher phenols, P, K, Ca, Zn and Cu contents in resistant genotype were also observed by Sindhan *et al.* (1988).

Jain and Chakarvarti (1988b) studied changes in transpiration rate due to bacterial infection and observed that rate of transpiration increased in blight infected leaves, which continued with increase in diseased leaf area. Reduction in amino acid contents like methonine, proline and phenylalanine after blight infection have also been reported (Singh and Swarup, 1984).

Light and nitrogen level in controlled environment affects the susceptibility of the host (Orellana and Thomas, 1968). However, under natural conditions of infection, application of 20-40 kg N/ha, which maintains the normal vigour of plant also increased the intensity of blight, though further increment in N dosages did not affect disease severity (Gandhi and Chand, 1985a). Use of healthy seed can effectively control the disease. Srivastava and Rao (1963b) have shown that seeds from summer grown clusterbean are free from infection. Hot water treatment at 56°C for 10 min. also eradicated the seed borne inoculum (Singh and Swarup, 1986). Seed treatment with streptomycin improved emergence of infected seeds and reduced blight intensity (Lodha, 1984; Sowell, 1965c). Foliar sprays with streptocycline (100 ppm) at 5th and 7th week after sowing reduced secondary spread of the disease and increased the seed yield (Lodha and Anantharam, 1993). This spary schedule was also cost effective (Table 1). Biological control to the tune of 58-80% was also achieved by spraying clusterbean crop with a white sterile fungus, *Penicillium citrinum*, *Aspergillus niger* and a bacterium *Bacillus subtilis* (Hooda *et al.*, 1995; Srivastava and Rao ed. 1963a, reported that glabrous varieties and some of pubescent branched types show considerable tolerance to the disease. Certain resistance sources have been identified and given below:

Resistant lines/varieties	Reference
IC 922 1/P3, PLG 74, PLG 191 and PLG 453	Mittal et al., (1977)
HFG 380, 522W, 590, 647, 648, 676	Gandhi et al. (1978a)
G85, G 102, G 126, G 225	Chahal et al. (1980)
GP 380, GP 508 Po GP 509B	Karwarsra and Chand (1982)
HG 75	Lodha (1984)

Table 1: Efficacy of different streptocycline spray schedules on disease intensity, yield and yield components of clusterbean

Time of Spray (week)	Per cent disease index (PDI)	Plant height (cm)	Pods plant^{-1}	Yield plant^{-1} (g)	Grain Yield (kg ha^{-1})
4	36.8 (37.3)[a]	83	45	7.3	841
5	35.2 (36.2)	77	47	9.2	845
4, 6	18.5 (25.3)	81	48	7.6	870
5, 7	16.6 (24.0)	85	52	9.8	992
7, 9	23.5 (28.9)	79	42	7.6	737
4, 6, 8	16.8 (24.2)	82	50	9.5	947
5, 7, 9	13.3 (21.3)	83	54	9.9	1023
Control	63.1 (52.6)	62	32	5.3	510
CD 5%	10.1	10.2	5.3	2.1	163

a Angular transformed values

2. Bacterial Leaf Spot

Another bacterial disease caused by *Pseduomonas cyamopsicola* was first time reported from India (Rangaswami and Gowda, 1963). Similar type of disease was also reported from Australia and U.S.A. (Orellana *et al.*, 1965). Initially minute water soaked lesions appear on leaves that soon turn light brown and become irregular in shape. Several such spots can be seen on each leaflet that coaleses to produce large patches with the advancement of disease. The leaves shrivel and dry up from tip downwards and ultimately drop off prematurely. In situation of severe infection extensive defoliation can be seen resulting in stunted plant growth and poor yield.

Another species, *P. syringae* has been reported from U.S.A. that produces lesions 1-5 mm in diameter, round to irregular with dark borders and light brown centres surrounded by chlorotic halos (Orellana, 1967a). It has been observed that both the pathogens (*P. syringae* and *X. cyamopsidis*) do not inhibit each other when grown in nutrient agar and there is no pronounced inhibition of diseases symptoms on leaves, when inoculated with a mixture of both the pathogens. The two organisms thus, can co-exist on this host in nature (Orellana, 1967a).

Viral Diseases

Guar serves as a local lesion host for a number of viruses. These include Beans top necrosis virus (Le Beau, 1947), Rose mosaic vines (Fulton, 1948), Sunhemp mosaic virus (Raychaudhry *et al.,* 1962), Strawberry mottle virus (Harris *et al.* 1986), Tobacco necrosis virus (Fulton, 1950), Tobacco ring spot virus and Top necrosis virus (Chester and Copper, 1944), Tobacco mosaic virus (Sekar and Sulochana, 1986) and Tobacco streak virus (Fulton, 1948).

Top necrosis considered to be a lethal disease in guar is caused by Tobacco ring spot virus (Chester and Copper, 1944). Symptoms include abscission of young leaves, terminal necrosis, necrotic stem lesions, yellowing and abscission of older leaves, marked stunting of growth and ultimate death.

This virus is sap transmissible to bean, lima bean, soybean, cow pea, mung bean, tobacco, husk tomato, garden pea and sesame (Chester and Copper, 1944; Orellana, 1966). The disease was also reported from India (Verma *et al.* 1962). It was observed tht guar selections were more susceptible to guar strain of TRSV than to tobacco strain (Orellana, 1967b). Only the wild *Cyamopsis senegalensis* and *C. serrata* were resistant to TRSV-g, thus, hybridization with *C. tetragonoloba* might be useful.

Mosaic disease of guar was reported from Hyderabad (Vani *et al.,* 1986). The disease was characterized by typical mosaic mottling and crinkling of leaves with moderate stunting, and phytomonas accumulation in leaves, thus whole plant reduced due to viral infection (Begum and Madhusudan, 1989). The virus is sap transmissible.

OTHER ROOT DISEASES

Apart from dry root rot, several other soil borne pathogens are responsible for certain root rots and wilts. Smith (1945) reported that lucerne root rot fungus *Rhizoctonia solani* was injurious to guar root also. In Uttar Pradesh, root rot caused by *R. solani, Fusarium caeruleum, Sclerotium rolfsii* and

Neocomospora vasinfecta are very destructive (Singh, 1951). *F caeruleum* caused wilting in 78% of the plants, while *S. rolfsii* and *N. vasinfecta* are weak parasites. Another pathogen *Ozonium taxanum var. parasiticum* was reported to cause wilt as well as leaf spot also (Mishra, 1963).

The incidence of *R. solani* and *F. caeruleum* was considerably reduced when guar was grown with sorghum, cowpea and moth in various combinations (Singh, 1954). These fungi are equally severe throughout moisture range of 15-16% and at temperature range of 28-30°C (Singh and Singh 1956). Improved plant health resulted from increased farmyard manuring, reduced root rot and wilt incidence in mature plants (Singh and Singh, 1957). The incidence of *S. rolfsii* increased with soil-moisture stress and acidic (pH 5.6-6.6) conditions (Singh and Gandhi, 1991), however, reduced with the addition of humus (Mathur, 1962). Liming of the soil and heavy manuring (compost) have also been found to reduce *S. rolfsii*.

Physiology of *F. coerulevum* has also been studied (Satyavir and Grewal, 1972; 1973). The dextrose aspargine phosphate solution gives maximum growth. Among carbon and nitrogen sources, maltose and DL-aspartic acid supported maximum growth of fungus.

SEED MYCOFLORA

A large number of fungi have been reported as external and internal contaminants from guar seeds. These can cause seed rot and often may serve primary sources of inoculum for certain diseases. Suryanarayana and Bhombe (1961) have reported *Alternaria, Aspergillus* and *Fusarium* as external and *Chaetomium* and *Gloeosporium* as internal contaminants of guar seeds. However, in subsequent studies, among internal seed borne fungi *Aspergillus* was also found to be common whereas, *Alternaria, Helminthosporium, Fusarium, Gliocladium, Phoma, Pleospora* and *Schizophylium* were less common (Singh and Chouhan, 1973). Karwasra and Singh (1982) isolated 9 fungi along with *Xanthomonas campestris PV. Cymopsidis* from seeds of guar. Among seed dressers, Agrosan GN,Captan, Zineb and Mancozeb were found most effective in reducing seed borne infection. Dwivedi and Dubey (1992) isolated 19 and 17 fungal species from unsterilized and sterilized stored seeds, respectively. *A. flavus, A. fumigatus* and *A. niger* were found predominate. *Aspergillus flavus* is the most dominant species. It inhibits root formation giving rise to stubby roots without any hair and with light brown stem (Singh and Chouhan 1973). Pathogenicity of *A. flavus* was determined in terms of its ability to cause seedling symptoms and abnormalities (Bhatia *et al.*, 1995). It was observed that it mainly resides in the seed coat, endosperm and spaces between endosperm and cotyledons. Essential oils isolated from seeds of *Trachyspermum amai, Cuminum cyminum, Carum carvi, Daucus carota* and from leaves of *Anethum graveolens* exhibited antifungal activity against *A. flavus*. Seed dressing fungicides induced germination and vigour of seedlings, however, aureofungin and PCNB were found more effective for the control of pre-emergence mortality.

FUTURE THRUST

In the last two decades, guar has assumed importance as a commercial crop in arid and semi-arid regions of India. Pearl millet was traditionally given more attention on the early onset of monsoon but in recent years due to high price of seed, growers have paid equal attention to guar also. Management of diseases to maximize productivity has also received equal priority in these years.

Concerted efforts at the level of policy planners, research and extension are needed to increase productivity of guar in rainfed areas. Disease management strategies require a basic information on the losses caused by important diseases. Efforts have to be made to identify durable multiple disease resistant sources against bacterial blight, powdery mildew and dry root rot, after rigorous testing at hot spots. Research efforts are required to integrate resistant genotypes, moisture conservation practices that reduce disease incidence, soil amendments, use of bio-control agents, efficient rhizobium strains, etc. Disease free areas should be identified for the production of healthy seeds.

REFERENCES

Agnihotri, J.P. and Prasad, N. 1966. Effect of growth regulating substances on growth of *Collectotricum capsici f. cyamopsicola* Prasad. *Labdev. J. Sci. Tech.* 4: 260-261.

Akem, C. and Lodha, S. 1997. Integrated management of soil-borne diseases of legumes in dry regions. In *Proceed. Indian Phytopath.Soc. – Golden Jubilee*, held at New Delhi Nov. 10-15, 508-511pp.

Arya, H.C. 1956. On a new leaf spot disease of gawar (*Cyamopsis tetragonoloba* Taub.) caused by *Myrothecium rosidum* Tode Ex.Fr. 9: 174-181.

Arya, H.C. 1959. On the occurrence of physiological strains in *Myrothecium rosidum*. *Indian Phytopath.* 12: 164-167.

Ayub, M., Mirza, J.H. and Shakir, A.S. 1995. Physiological studies of *Alternaria cyamopsidis* Rang and Rao causing leaf blight of guar (*Cyamopsis tetragonoloba* Taub.). *Pakistan J. Phytopath.* 7: 41-43.

Begum, H. and Madhusudan, T., 1989. Effect of Mosaic virus infection on growth and yield of clusterbean. *Indian J. Mycological Res.* 27: 191-194.

Bhatia, A., Singh, T. and Singh, D. 1995. Invasion of *Aspergillus flavus* in clusterbean seeds and its pathogenic behaviour. *Seed Res.* 23: 50-52.

Butler, E.J., 1918. Fungi and Diseases in Plants. Thacker Spink & Co., Calcutta.

Chahal, S.S., Sahoo, M.S. and Gill, K.S., 1980. Source of resistance against bacterial blight of guar (*Cyamopsis tetragonoloba*). *Indian J. Mycol. Pl. Pathol.* 9: 103-104.

Chand, J.N. and Gandhi, S.K. 1978. Diseases of guar and their control. *Forage Res.* 40: 49-66.

Chand, J.N. and Verma, P.S. 1968a. A leaf spot disease of *Cyamopsis tetragonoloba* caused by *Curvularia lunata* in India. *Indian Phytopath.* 21: 239-240.

Chand, J.N. and Verma, P.S. 1968b. Occurrence of a new Alternaria leaf spot of clusterbeans (*C. tetragonoloba*) in India. *Plant Dis. Reptr.* 52: 145-147.

Chandra, U. and Saxena, S.K. 1990. Occurrence of *Sphaerotheca fuliginea* (Sschelecht ex. Fr.) Poll on guar bean (*Cyamopsis tetragonoloba* L. Taub.) in and around Aligarh. *Nation. Acad. Sci. Letters* 13: 261-262.

Chestor, K.S. and Cooper, W.E., 1944. A lethal virus of guar (*Cyamopsis psoralioides* DC). (Absts). *Phytopathol.* 34: 998.

Desai, M.V. and Prasad, N., 1955. A new *Collectotrichum* from India. *Indian Phytopath.* 8: 52-57.

Dwivedi, S.K. and Dubey, N.K. 1992. Mycoflora associated with seeds of *Cyamopsis tetragonoloba* L. (Taub.) *J. Mycopathol. Res.* 30: 153-156.

Fulton, R.W. 1948. Hosts of Tobacco streak virus. *Phytopathol.* 38: 421-428.

Fulton, R.W. 1950. Variants of Tobacco necrosis virus in Wisconsin. *Phytopathol.* 40: 298-305.

Gandhi, S.K. and Chand, J.N. 1985a. Yield losses in guar due to bacterial blight caused by *Xanthomonas campestris pv. Cyamopsidis*. *Indian Phytopath*. 38: 516-519.

Gandhi, S.K. and Chand, J.N. 1985b. Effect of nitrogen and phosphorous on the bacterial blight of clusterbean. *Indian J. agric. Sci.* 55: 376-377.

Gandhi, S.K. Chand, J.N. and Saini, M.L. 1978a. Reaction of genetic stock of guar (*Cyamopsis tetragonoloba* (L.) Taub.) to bacterial blight (*Xanthomonas Cyamopsidis*) and its incidence in relation to date of sowing. *Forage Res.* 4: 159-162.

Gandhi, S.K. Saini, M.L. and Jhorar, B.S. 1978b. Screening of clusterbean genotypes for resistance to leaf spot caused by *Alternaria cyamopsidis*. *Forage Res.* 4: 169-171.

Gupta, D.K. 1978. Yield losses in guar (*Cyamopsis tetragonoloba*). *Indian J. Mycol. Pl. Pathol.* 8: 27.

Harris, D.C. Funnell, P.S., Davies, M.K. Talboys, P.W. Cardon, J.A. Simpson, D.W. Adams, A.N. and Blake, V.S.A. 1986. Diseases of strawberries. *Report of East Maling Res. Stn.*, 121 pp.

Hooda, I., Parashar, R.H. and Sindhan, G.S. 1995. Assessment of per cent *Xanthomonas campestris pv. Cyamopsidis* infection in clusterbean seed raised from crop sprayed and seed treated with antagonists for controlling bacterial blight infection. *Plant Dis. Res.* 10: 70-73.

Jain, J.P. and Chakarvarti, B.P. 1988a. Changes in transpiration rate of guar leaves due to infection with *Xanthomonas campestris pv. Cyamopsidis*. *Indian J. Mycol. Pl. Pathol.* 18: 58-59.

Jain, J.P. and Chakarvarti, B.P. 1988b. A comparison of the methods of detection of Bacterium, *Xanthomonas campestris pv. Cyamopsidis* in guar seeds. *Indian J. Mycol. Pl. Pathol.* 18: 107-108.

Karwasra, S.S and Singh, M. 1982. Seed mycoflora of clusterbean in Haryana and their control by seed treatment. *Indian Phytopath.* 35: 501-502.

Karwasra, S.S. and Chand, J.N. 1982. Screening of genetic stock of guar (*Cyamopsis tetragonoloba*) to bacterial blight (*Xanthomonas campestris pv. Cyamopsidis*). *Indian J. Mycol. Pl. Pathol.* 12: 229.

Khan, S.A. 1958. Powdery mildew of Tandojam. *Pak. J. Sci. Res.* 10: 82.

Kernhamp, M.F. and Hemerick, G.A. 1953. The relation of *Ascochyta* to Alfalfa seed production in Minnesota. *Phytopathol.* 43: 378-383.

Kothari, K.L. and Bhatnagar, M.K. 1966. Evaluation of fungicides against crop diseases. 1. On *Colletotrichum capsici* causing blight of guar. *Indian Phytopath.* 19: 116-117.

Lodha, S. 1984. Varietal resistance and evaluation of seed dressers against bacterial blight of guar (*Cyamopsis tetragonoloba*). *Indian Phytopath.* 37: 438-440.

Lodha, S. 1995. Soil solarization, summer irrigation and amendments for the control of *Fusarium oxysporum f. sp. cumini* and *Macrophomina phaseolina* in arid soils. Crop Prot. 14: 215-219.

Lodha, S. 1996. Influence of moisture conservation techniques on *Macrophomina phaseolina* population, dry root rot and yield of clusterbean. *Indian Phytopath.* 49: 342-349.

Lodha, S. and Anantha Ram K. 1993. Effectiveness of streptocycline spray schedules on the bacterial blight intensity and yield of clusterbean. *Ann. Arid Zone.* 32: 237-240.

Lodha, S. and Gupta, G.K. 1981. The relationship between bacterial blight and loss in yield components of guar (Abstr). In *Proc. Third Inter. Symp. on Pl. Pathol.*, held at New Delhi, Dec. 13-17, 94-95.

Lodha, S., Mali, P.C. and Burman, U. 1993. Development of bacterial blight and changes in biochemical components in the resistant and susceptible genotypes of clusterbean. *Indian Phytopath.* 46: 354-359.

Lodha, S., Mathur, B.K. and Solanki, K.R. 1990. Factors influencing population dynamics of *Macrophomina phaseolina* in arid soil. *Plant Soil.* 125: 75-80.

Lodha, S. and Solanki, K.R. 1993. Inheritance of dry root-rot resistance in clusterbean. *Indian Phytopathol.* 45: 430-433.

Lodha, S., Sharma, S.K. and Aggarwal, R.K. 1997. Solarization and natural heating of irrigated soil amended with cruciferous residue for improved control of *Macrophomina phaseolina*. *Plant Pathol.* 46: 186-190.

Mali, P.C., Vyas, S.P. and Lodha, S. 1989. Biochemical components of clusterbean genotypes in relation to bacterial blight. *Indian Phytopath.* 42: 559-561.

Mathur, R.L. and Batrice, M. 1971. Bioassay of wettable sulphur compounds against *Oidiopsis taurica* (Lev.) Saln. causing powdry mildew of *Cyamopsis psoraloides*. *Labdev J. Sci. Tech.* 9B: 76-77.

Mathur, R.L., Mathur, B.N. and Sharma, B.S. 1972. Relative efficacy of fungicides for the control of *Alternaria cyamopsidis* causing leaf spot of guar (*Cyamopsis tetragonoloba*). *Indian J. Mycol. Pl. Pathol.* 2: 80-81.

Mathur, S.B. 1962. Effect of soil moisture on root rot of guar

(*Cyamopsis psoraloides* DC.) and wilt of gram (*Cicer arietinum* L.) caused by *Sclerotium rolfsii Sacc. Agra Univ. J. Res.* 11: 295-302.

Mishra, J.N. 1963. Ozonium wilt of guar and *Cucurbit* plants in Bihar. *Nature*. 172: 209-210.

Mittal, S.P., Dabas, B.S., Thomas, T.A. and Chopra, D.P. 1977. Assessment of genetic stocks of guar (*Cyamopsis tetragonoloba* (L.) Taub.). *Proc. Workshop on Guar Res. & Prod.*, Central Arid Zone Research Institute, Jodhpur, 58-68.

Orellana, R.B., 1966. A new occurrence of Tobacco ring spot of guar in United States, *Plant Dis. Reptr.* 50: 7-10.

Orellana, R.G. 1967a. Leaf spot of guar caused by *Pseudomonas syringae* in the United States. *Plant Dis. Reptr.* 51: 182-184.

Orellana, R.G. 1967b. Reaction of guar to strains of Tobacco ring spot virus. *Phytopathol.* 57: 791-792.

Orellana, R.G., Thomas, C.A. and Kinman, M.L. 1965. A bacterial blight of guar in United States. *FAO Plant Prot. Bill.* 13: 9-13.

Orellana, R.G. and Simmons, E.G. 1966. Alternaria blight of guar in the United States. *Mycopathol. Mycol. Appl.* 29: 129-133.

Orellana, R.G. and Thomas, C.A. 1968. Light and nitrogen affect reaction of guar to bacterial blight caused by *Xanthomonas cyamopsidis*. *Phytopathol.* 58: 250-251

Orellana, R.G. and Weber, D.F. 1971. Immuno-diffusion analysis of isolates of *Xanthomonas cyamopsidis*. *Appl. Microbiol.* 22: 622-624.

Parashar, R.D., Sharma, D.D.K. and Sindhan, G.S. 1994. Variability in *Xanthomonas campestris pv. Cyamopsidis* the causal agent of bacterial blight of guar. *Plant Dis. Res.* 9: 127-133.

Patel, M.K., Dhande, G.W. and Kulakarni, Y.S. 1953. Bacterial leaf spot of *Cyamopsis tetragonoloba* (L.) Taub. *Curr. Sci.* 22: 183.

Patel, A.J. and Patel, M.K. 1958. A new bacterial blight of *Cyamopsis tetragonoloba* (L.) Taub. *Curr. Sci.* 27: 258-259.

Prasad, N. 1945. Studies on the root rot of cotton in Sind II. Relation of root rot of cotton with root rot of other crops. *Indian J. agric. Sci.* 14: 388-391.

Prasad, N. and Desai, M.V. 1952. Fusarium blight of clusterbeans. *Curr. Sci.* 21: 17-18.

Praveen Kumar, Singh, Y.V., Lodha, S. and Aggarwal, R.K. 1998. Efficient management of resources for sustainable crop production in arid region. *Res. Bull.*, CAZRI, Jodhpur, 72pp.

Rangaswami, G and Rao, V. 1957. Alternaria Blight of clusterbean.

Indian Phytopath. 10: 18-25.

Rangaswami, G. and Gowda, S.S. 1963. On some bacterial diseases of ornamentals and vegetables in Madras State. *Indian Phytopath.* 16: 74-85.

Rao, P.G. and Rao, D.K. 1956. An anthracnose disease on *Cyamopsis tetragonoloba* Taub. *Sci. Cult.* 21: 457-458.

Ratnam, C.V., Pandit, S.V. and Rao, K.C. 1985. Evaluation of some systemic and non-systemic fungicides against powdery mildew of clusterbean. *Pesticides* 19: 36-37.

Raychaudhuri, S.P., Nariaini, T.R. and Das, C.R. 1962. *Cyamopsis tetragonoloba* (L.) Taub.- a local lesion host of sunhemp mosaic virus. *Indian Phytopath.* 15: 79-80.

Reddy, M.S. and Rao, A.A. 1971. Benlate: A systemic fungicide highly effective against Powdery mildews. *Indian Phytopath.* 24: 196-197.

Satyavir and Grewal, J.S. 1972. Evaluation of fungicides against *Fusarium caeruleum* the casual organism of guar wilt. *Indian Phytopath.* 25: 72-75.

Satyavir and Grewal, J.S. 1973. Studies on the physiology of *Fusarium caeruleum* the casual organism of guar wilt. *Indian Phytopath.* 26: 62-75.

Sekar, R. and Sulochana, C.B. 1986. Studies on cross protection behaviour on TMV strains in a local lesion host. *Indian J. Pl. Pathol.* 4: 33-39.

Sharma, S.R. 1983. Effect of fungicides on the development of *Alternaria* blight and yield of clusterbean. *Indian J. agric. Sci* 53: 932-935.

Sharma, S.R. 1984. Effect of different cropping seasons and tridemorph sprays on powdery mildew and yield of clusterbean. *Pflanzen Z.* 2: 81-89.

Sharma, S.K., Aggarwal, R.K. and Lodha, S. 1995. Population changes of *Macrophomina phaseolina and Fusarium oxysporum f. sp. cumini* in oil-cake and crop residue amended sandy soils. *App. Soil Eco.* 3: 281-284.

Shivanna, M.B. and Shetty, H.S. 1986. Myrothecium pod spot of clusterbean and its significance. *Curr. Sci.* 55: 574-576.

Sidhan, G.S., Parashar, R.H. and Hooda, I. 1988. Role of phenolic compounds and mineral elements in resistance of some guar cultivars to bacterial blight. *Plant Dis. Res.* 3: 48-49.

Singh, R.S. 1951. Root rot of guar. *Sci. Cult.* 17: 131-134.

Singh, R.S. 1954. Effect of mixed cropping on the incidence of root rot, wilt and associated disease of *Cyampsis psoraloides*. *Agra Univ. J. Res.* 3: 359-373.

Singh, B. and Singh R.S. 1956. Temperature and moisture relations of the fungi causing seedling rot, root and wilt of *Cyamopsis psoraloides* DC 1. Effect of temperature on the growth of fungi in artificial media 2. Effect of soil moisture on mortality under controlled conditions. *Agra Univ. J. Res.* 5: 135-141.

Singh, B. and Singh, R.S. 1957. Incidence of seedling rot, root-rot and wilt of *Cyamopsis psoraloides* under manured and unmanured conditions. *Agra Univ. J. Res.* 6: 7-14.

Singh, I. and Chouhan, J.S. 1973. Seed mycoflora of 'guar' (*Cyamopsis tetragonoloba*) and their effect on germination and growth of seedlings. *Indian J. Mycol. Pl. Pathol.* 3: 86-92.

Singh, S.D. and Prasada, R. 1973. Studies on physiology and control of *Alternaria cyamopsidis* the incitant of blight disease of guar. *Indian J. Mycol. Pl. Pathol.* 3: 33-39.

Singh, G., Gupta, R.B.L. and Dalela, G.G. 1974. Efficacy of fungicides and varietal resistance of clusterbean (*Cyamopsis tetragonoloba*) against leaf spot disease caused by *Curvularia lunata*. *Indian Phytopath.* 27: 234-236.

Singh, J. and Swarup, J. 1984. Quantitative estimation of amino acids in clusterbean leaves infested with *Xanthomonas campestris pv. Cyamopsidis*. *Indian J. Mycol. Pl. Pathol.* 14: 80-81.

Singh, J., Swarup, J. and Singh, J. 1988. Survival and virulence of *Xanthomonas campestris pv. Cyamopsidis* in plant debris, seed, soil and *in vitro*. *Indian J. Pl. Path.* 6: 93-95.

Singh, J. and Swarup, J. 1986. Control of bacterial blight of clusterbean by seed treatment and foliar spray. *Pesticides.* 20: 23-24.

Singh, J. and Swarup, J. 1987. Epidemiology of bacterial blight of clusterbean caused by *Xanthomonas campestris pv. Cyamopsidis*. *Farm Sci. J.* 2: 91-92.

Singh, R.P. and Gandhi, S.K. 1991. Effect of soil pH and temperature on seedling mortality of guar caused by *Sclerotium rolfsii* and its fungicidal control. *Indian Phytopath.* 44: 360-365.

Smith, O.F. 1945. Parasitism of *R. solani* from alfalfa. *Phytopathol.* 35: 832-834.

Solanki, J.S. and Singh, R.R. 1976. Note on the fungicidal control of powdery mildew of clusterbean. *Indian J. agric. Sci.* 46: 241-243.

Sowell, G. 1965a. Alternaria leaf spot of guar. *Plant Dis. Reptr.* 49: 605-607.

Sowell, G. 1965c. The effect of seed treatment on seed borne pathogen of guar. *Plant Dis. Reptr.* 49: 895-897.

Srivastava, D.N. and Rao, Y.P. 1963a. Bacterial blight of guar. *Indian Phytopath.* 16: 69-73.

Srivastava, D.N. and Rao, Y.P. 1963b. Seed transmission and control of bacterial blight of guar (*Cyamopsis tetragonoloba* (L.) Taub.). *Indian Phytopath.* 16: 390-391.

Streets, R.B. 1948. Diseases of guar (*Cyamopsis psoraloides*). *Phytopathol.* (Abstr.) 38: 918.

Suryanarayana, D. and Bhombe, B.D. 1961. Studies on fungal flora of some vegetable seeds. *Indian Phytopath.* 14: 30-41.

Vani, V., Madhusudan, T. and Rao, R.D.V.J.P. 1986. A new mosaic disease of clusterbean. *Indian J. Virol.*, 2: 222-230.

Vasudeva, R.S. 1960. Report of Division of Mycology and Plant Pathology. *Sci. Rep. Agric. Res. Inst.*, N. Delhi. 1957-1958, pp 111-130.

Venkatarayan, S.B. 1946. A natural fungous parasite of Powdery mildew on *C. psoraloides* DC. *Curr. Sci.* 15: 319.

Verma, G.S., Varma, J.P. and Saxena, P.N. 1962. Top necrosis of *Cyamopsis tetragonoloba* (L.) Taub. *Proc. Indian Acad. Sci.* 32: 287-292.

Westhuizen, G.C.A., Van-der and Beck, B.D.A. 1987. Occurrence of Alternaria blight of guar in South Africa. *Phytophylact.* 19: 461-465.

CHAPTER - 9

INSECT PESTS

R.K. Arora and R.K. Kashyap

Several insect and mite pests have been reported to be associated with clusterbean crop from all over the world (Dhaliwal, 1980; Singh *et al.*, 1981; Butani and Jotwani, 1984; Whistler and Hymowitz, 1979). Insects belonging to 41 genera from 16 families and 8 orders are associated with this crop the world over (Table-1). An attempt has been made to review the available information for each insect species individually.

GALL MIDGES

Asphondylia cyamopsil n. sp.

The first record of attack by an *Asphondylia* sp. which produced galls on garden guar was reported from India by Rao (1917). During 1967, the commercial crop of guar suffered heavy infestation due to gall midge attack *Asphondylia cyamopsii* n sp. By inducing pod gall formation (Patel and Chari, 1970).

Damage

The infestation is caused by maggots which result in the production of monothalmus galls. Bud galls become bigger, hard and do not produce flowers. Due to the damage, elongation of cells both in endocarp and epicarp (fruit wall) results in their expansion (Chari, 1971). Because of the elongation of the cells in the tissues surrounding carpellary cavity, stigma of the flowers does not develop normally and remains like a hook. This hook like structure of the stigma is a peculiar diagnostic character of gall midge infestation. There is marked decrease in chlorophyll content in the infested galls which appear pale green till the emergence of the adult. The young pods when attacked results in arrested pod maturation and seed development. The galled pods turn brown and dry in about 18 to 24 days after the emergence of the midge.

Table 1. Insect and mite pests of clusterbean (*Cyamopsis tetragonoloba* (L.) Taub)

Scientific name	Common name	Family	Reference
Order: Diptera			
Ashondylia spp.	Gall midge	Cecidomyiidae	Patel and Chari (1970)
Contarinia texana (Felt.)	Guar midge	Cecidomyiidae	Rogers (1972)
Phytomyza atricormis (Meign)	Pealeaf miner	Agromyzidae	Butani (1980)
Order: Homoptera			
Aphis spp.	Aphid	Aphididae	Singh *et al.* (1981)
Aphis craccivora (Koch)	Cowpea aphid	Aphididae	Dennis (1976)
Acyrthosiphon pisum (Harris)	Pea aphid	Aphididae	Dennis (1976)
Order : Hemiptera			
Acreatagallia uhleri (Van Duzee)	Leaf hopper	Cicadellidae	Dennis (1976)
Empoasca sp.	Cotton jassid	Cicadellidae	Abbas and Afzal (1946)
E. motti Pruthi	Leaf hopper	Cicadellidae	Dhaliwal (1980)
E. keri	Leaf hopper	Cicadellidae	Buti *et al.* (1982)
Spissistilus festinus (Say)	Three cornered alfa-alfa hopper	Cicadellidae	Streets (1948)
Coptosoma cribraria (Fab.)	Lab-lab bug	Corimelaenidae	Butani and Jotwani (1984)
C. nazirae (Atkinson)	Stink bug	Thyrocoridae	
Cyclopelta siccifolia (Westwood)	Stink bug	Pentatomidae	Butani and Jotwani (1984)
Trialeurodes abutilonea (Haldeman)	Whitefly	Aleyrodiaidae	Dennis (1976)
Bemisia tabaci Genn.	Cotton whitefly	Aleyrodiaidae	Singh *et al.* (1996)
Order: Thysanoptera			
Ayyaria chaetophora Karney	Thrips	Thripidae	Dennis (1976)

Frankliniella shultzei (Trybom) (=dampfi Priesner)		Thripidae	Butani and Jotwani (1984)
Megleurothrips usitatus (Bangall)	Blossom thrips	Thripidae	Butani and Jotwani (1984)
Taeniothrips distalis (Karny)	Blossom thrips	Thripidae	Butani and Jotwani (1984)
Order : Lepidoptera			
Amsacta moorei Butler	Redhairy caterpillar	Arctiidae	Srivastava and Goel (1962)
Diacrisia oblique	Biharhairy Caterpillar	Arctiidae	Dhaliwal (1980)
Heliothis zea (Boddie)	Corn earworm	Noctuidae	Dennis (1976)
H. armigera Hubner	Gram caterpillar	Noctuidae	Singh *et al.* (1981)
Tethrohynchus angustiorata Grote	Looper	Noctuidae	Dennis (1976)
Languiria spp.	Stem borer	Crambidae	Dennis (1976)
Platynota stultana Walsingham	Leaf roller	Crambidae	Dennis (1976)
Dichomeris ianthus Meyr.	Leaf perforator	Gelechidae	Dhaliwal (1980), Singh *et al.* (1982)
Order : Coleoptera			
Alcidodes bubo (Fab.)	Agathi stem Weevil	Cruculionidae	Subramanian (1959)
A. fabricii (Fab.)	Striped sweet Potato weevil	Cruculionidae	Subramanian (1959)
A. collaris Pascoe	Potato weevil	Cruculionidae	Subramanian (1959)
A. pictus Pascoe	Potato weevil	Cruculionidae	Subramanian (1959)
Apion leopardus olivier	Flea beetle	Cruculionidae	Subramanian (1959)
Blosyrus inaequalis Bohemann	Bean Weevil		Butani and Jotwani (1984)
Cyrtozemia cognata Marsh.	Leafeating weevil	Cruculionidae	
C. dispar Marshall	Leafeating weevil	Cruculionidae	
Desmidophorus Spp.	Leafeating weevil	Cruculionidae	
Myllocernus maculosus Desbrochers	Cotton grey weevil	Cruculionidae	
Conicobruchus albopubens	Seed beetle	Bruchidae	Arora (1977)
Order : Orthoptera			
Chrotogonus trachypterus	Surface Grasshopper	Acrididae	Singh *et al.* (1981)
Order : Acacrina			
Polyphagotarsonemus latus (Banks)	Yellow mite	Tarsonemidae	Singh *et al.* (1981)

Life cycle

Adults of *A. cyamopsii* are dusky drab in colour, measure 3-4 mm long with a wing expanse of 5.5 mm and are strong fliers. Females have a protractile ovipositor and deposit eggs in the ovary of the flower buds and pods. On hatching, the maggot starts feeding on the ovules. Full grown maggots are yellowish in colour, 1.0-1.25 mm wide and 3.25-3.96 mm long. Pupation occurs within the galls. Pupae exceed 3 mm in length and possess well developed cutting organs. This structure presumably serves to cut the gall epidermis prior to imago emergence (Abbas, 1967). Pupal exuvae are often left protruding from the exit hole of the gall when adult emerges.

Management

The maggots of gall midges are parasitized by *Eurytoma denctipectus* Gohan and *E.nesiotes*. Growing of nectar bearing plants, viz. Corriander, alfa-alfa, etc. as companion plants to provide suitable habitat for greater survival of parasitoids merits serious thinking (Srinivasan, 1994).

Manual removal and destruction of the infested buds and avoidance of late sowing help to suppress the incidence of the pest (Rai, 1976). Spraying with carbaryl 0.15% or phosphamidon 0.03% at flower initiation stage (Singh and Yadava, 1985) or 0.05% monocrotopos (Krishna Kumar and Sardana, 1988) provide effective control of the pest.

HAIRY CATERPILLARS

Red hairy caterpillar

In India, *Amsacta moorei* Buttler (= *A. albistriga* Wlk.) has been reported from sandy areas, particularly in the states of Rajasthan, Haryana, Gujarat and Punjab (Saini, 1993). The sandy texture of the soil in these areas provides favourable sites for pupation as well as moth emergence. In Rajasthan, outbreak of the pest in certain years was ascribed to the early, heavy and frequent rains supported by sandy and sandy loam soils (Yadava *et al.,* 1966).

It is a polyphagous pest and is serious particularly during *kharif* season.

Damage

The damage is caused by the larval stages. Neonates feed gregariously and their feeding propensity increases with the age. As they grow older, they disperse. The catterpillars feed voraciously and indiscriminately practically finishing the vegetation in the fields where the initial egg laying took place. During years of out breaks, usually fields have to be resown.

Life cycle

The adults are generally medium sized moths. The average body length and wing expanse of male (18.75 mm; 40.40 mm) and female (20.80 mm; 42.25 mm) varies depending on different host plants (Ahlawat *et al.,* 1993). The forewings are white with anterior marginal streaks and a red band on head. The hind wings are white with black spots. The anterior margin of the thorax and the entire abdomen are scarlet red. There are black bands and dots on the abdomen. In *A. albistriga*, the forewings have brownish streaks all over and yellowish streak along the anterior margin. Both species are found together.

The caterpillars are covered with numerous long hairs arising from the fleshy tubercles. Their colour varies from reddish amber or brownish-orange to olive green. The full grown larva have been reported to measure about 25 mm (Atwal and Dhaliwal, 1997), 50-70 mm (Kumar and Nigam, 1989) to 6 centimetres in different ecological conditions.

The pest is active from end of June to the middle of October. The moths emerge after the first few good showers of monsoon; lay spherical light yellowish eggs (measuring 0.72 x 0.69 mm in size) in batches of 700-850 each, mainly on the under surface of leaves. A single female may lay upto 1500 eggs during its life time. The incubation period varies from 3 to 5 days. The larvae pass through seven instars and complete their development in 15-23 days. The full grown larvae shed their hair, pupate in soil after making earthen cocoons at a depth of about 23 cm.

Management

Mechanical control measures viz., mass trapping, manual picking and deep ploughing have been found quite effective for controlling red hairy caterpillars. The mass trapping of adults in light traps during night has been reported to be highly effective in controlling *Amsacta* spp (Saini and Verma, 1991). The light traps placed just above a broad flat basin full of kerosenized water should be used following the first showers of the monsoon and this should be continued throughout the period of emergence of about one month. The young larvae feeding gregariously on leaves can be destroyed by plucking the leaves or cutting branches or by pulling out plants harbouring the larvae and subsequently destroying them or burying them underground. The grown up larvae can be destroyed by crushing them under feet or picking and putting them in kerosenized water. After the harvest of the crop, deep ploughings and if possible flooding the infested fields to kill the pupae present in the soil prevents the carry-over of the pest.

Chemical control measures viz., the use of sevimol 0.08 per cent against first, second and third instar, Sevimol 0.15 per cent against fourth and carbaryl 0.2 per cent against fifth instar larvae and chlorophyriphos 0.1 per cent to kill fourth and fifth instar larvae have been recommended. Quinalphos (0.05 per cent), monocrotophos (0.05 per cent) and triazophos (0.06%) have also been recommended for good control of sixth instar larvae (Saini, 1995).

Bihar Hairy Caterpillar *Silosoma (=Diacrisia) obliqua* **Walker**

It is a sporadic and polyphagous pest widely distributed throughout the Orient. *S.obliqua* has been reported to cause considerable damage to guar crop. It usually appears in the field when the crop is comparatively young (Singh *et al.*, 1981).

The early instars feed gregariously on leaves, soft portions of stems and branches. As a result, the chlorophyll of the leaves is removed. Older larvae migrate to other fields and may cause defoliation resulting in meshed appearance of the crop.

Life cycle

Adult moths have crimson or dull yellow head; thorax and under side of the body has black dots. Wings are pinkish buff in colour with numerous black spots. The moths measure about 40-50 mm across the wings. Full grown caterpillars are stout, 25 to 45 mm long and have 7 orange coloured broad transverse bands with tufts of long greyish hair all over the body which are dark at both the ends.

The pest is active from March-April and again from July to November. The adults emerge from the over wintering larvae in March and immediately start mating during night. The female lays eggs in clusters on ventral surface of leaves in parallel rows on 3 to 4 consecutive nights. A single female may lay 400-1600 eggs in its life time. Incubation period lasts for 4-13 days. The larvae undergo seven instars and the larval period varies from 28 to 56 days. The last instar larva spins a loose silken cocoon after which pupation takes place in plant debris or in soil. The pre-pupal and pupal stage lasts for 2 and 10 to 15 days, respectively (Atwal and Dhaliwal, 1997). The life cycle is completed in 5 weeks in summer but is extended upto 8 and 10 weeks during autumn and winter. There may be as many as 8 overlapping generations in a year. The pest remains in pupal diapause during summer (May-June) and also during winter (December-February).

Management

Mechanical measures viz., Collection and destroying the leaves bearing egg-masses promptly before the larvae hatch from the eggs help to prevent the population build up of Bihar hairy caterpillar. Neonates can either be destroyed by pulling out the infested plants and burying them underground or by dusting the crop with malathion 5 per cent dust @ 25 kg/ha or trichlorophon 5D @ 35 kg/ha.

In case of severe infestation of *Spilosoma oblique,* spraying 1.25 litres of endosulfan 35 EC or quinalphos 25 EC or 500 ml of dichlorvos 100 EC in 500 litres of water per ha is quiet effective against all stages of this pest.

Apanteles oblique Walker and *Eurytome* spp. have been recorded as larval parasites of this pest (Butani and Jotwani, 1984).

APHIDS

Aphids are polyphagous insect pests and have been reported to inflict serious damage to legumes world over (Chhabra *et al.,* 1993; Chhabra and Kooner, 1989). The aphids have been reported to damage young plants of guar in the early stages of the crop from July to September. It has been reported that aphids thrive best in damp weather and when moist wind blows (Singh *et al.,* 1981). Aphid cause damage either by lowering the plant vigour by sucking the cell sap or by transmitting viruses in various crops. *A. pisum* and *A. craccivora* have been reported to be associated with guar crop.

The Pea aphid (*Acyrthosiphon pisum*) is a sporadic pest primarily infesting pea crop.

A.pisum has been reported to transmit various types of viral diseases viz., alfa alfa mosaic virus, pea leaf roll virus, cowpea mild mottle virus, bean yellow mosaic virus in chickpea, and cowpea and lentil (COPR, 1981).

Life cycle

Adults are pear shaped, green yellow or pink in colour with long conspicuous cornicles. Both alate as well as apterous forms are present which are generally females. Males are rare. Winged as well as wingless males have been recorded from Europe and USA but not from India (Butani and Jotwani, 1984).

Reproduction is parthenogenetic and it takes about seven days to complete one generation. There are several overlapping generations in a year.

Bean or groundnut aphid: *Aphis craccivora (=A. leguminosae* Th) is another most common and destructive aphid species damaging clusterbean, Indian beans and other leguminous crops. It is widely distributed throughout the world.

A.craccivora has been reported as vector to transmit virus diseases viz., mosaic mottling virus, mild mottle virus, alfa-alfa mosaic virus, bean yellow mosaic virus, cowpea aphid bean mosaic virus (CAbMV), pea leaf roll virus and Iranian strain virus (Kaiser, 1979; COPR, 1981; Thotapilly and Rossel, 1985).

Life cycle

Detailed studies have been conducted on the bionomics of *A.craccivora* by Bakhetia and Sidhu (1977). The nymphs are wingless, dark and fairly rounded in body shape. Adults are black or dark brown and variable in size. They reproduce parthenogenetically and viviparously. A female may produce 8-30 young ones in life span of 10-12 days. However, it was reported that a single female was capable of producing 174.3 nymphs in 15 days. The nymphs undergo four moults to become adults in 5-8 days. The duration of each instar is usually one day. One day old apterous female begins to produce brood. A female could reproduce upto a maximum of 12 days. However, fecundity and development rates vary with host plant, soil fertility, soil moisture and temperature. Drought conditions are reported to be responsible for its multiplication and spread over large areas.

Management

Intercropping, preferably with non-leguminous crops, use of alternate hosts, manipulation in date of sowing and spacing help to achieve optimum control of two aphid species (Chhabra *et al.,* 1993; Prasad *et al.*, 1988; Lal *et al.,* 1989). Similar work needs to be taken up in clusterbean crop which has been reported to become susceptible to serious aphid infestation. In India, 24 predators and nine parasites have been found to be associated with these two aphid species.

Dimethoate, phosphamidon, monocrotophos, oxydemeton-methyl and malathion; spray formulation of methyl-o-demeton, dimethoate, dichlorvos, endosulfan, phosalane, cypermethrim and fenuvalerate and soil application of phorate have been reported to be effective against aphids. Khurana and Kaushik (1991) reported synthetic pyrethroids to be as effective as conventional insecticidal treatments against *A.craccivora.*

Effective control of *A.craccivora* has been obtained with the use of primicarb (a highly selective aphicide), dimethoate and thiometon and application of Lambda-Cyhalothrin @ 5.0-7.5 g a.i/ha (Pruszynski and Mrowesynski, 1990).

WHITE FLIES

The white flies are minor pests of clusterbeans. The species *Bemisia tabaci* (*gennadius = gossypiperda* Misra and Lamba) is widely distributed in tropics and subtropics and is known in India since 1905 as cotton whitefly (Husain and Khan, 1945). *Bemisia tabaci* (Genn.) is a polyphagous pest. In guar, the incidence of *B.tabaci* is increasing for the last 4-5 years.

Both nymphs and adults suck the sap from the ventral surface of the leaves. As a result of their feeding, the affected parts become yellowish, the leaves wrinkle and curl downwards and are ultimately shed. Besides the feeding damage, these insects also exude honeydew which favours the development of sooty mould which hinders the photosynthesis of the plant resulting in stunting growth. Not only the seed yield is considerably reduced but the quality of the fodder also deteriorates, if the crop is grown for fodder purpose. This whitefly also acts as a vector of leaf curl virus diseases causing severe losses. The pest is active during the dry season and its activity decreases after the onset of monsoons.

Life cycle

Eggs are 1.5 to 2.0 mm, pyriform or ovoid, light yellow in colour. The maximum number of eggs per female varies between 43 to 394. On hatching, the nymphs crawl a little, settle down on a succulent spot on the same leaf and never move again during that stage. Nymphs are oval, scale-like and greenish-white in colour. Adults are about one mm long, covered completely with a white waxy bloom. Wings are opaque and milky white in colour; wing expanse is 4 to 5 mm. Reproduction is mostly oviparous, though parthenogenesis also occurs. Incubation period is 3 to 5 days in summer extending upto 33 days in winter. Nymphal development takes 9 to 14 and 17 to 81 days in summer and winter, respectively. Fourth nymphal instar (commonly referred to a pupa (Gill, 1990) lasts for 2 to 8 days. A life cycle is completed in as little as 14 days to a prolonged duration upto 107 days. Adult longevity is 2 to 5 days in summer and 24 days in winter. There are about 12 overlapping generations in a year.

Management

Sixty clusterbean genotypes were screened to find out the source(s) of resistance against whitefly. None of them was found free from the attack of whitefly. However, five genotypes, viz., CH14-2, HG-75, HG-94, HG-258 and HGS-365 were found least preferred. RGC 1001 was recorded as highly susceptible. HG-75 and HG-258 has already been recommended by Central Variety Release Committee for general cultivation all over India (Singh *et al*., 1996).

Rao *et al.* (1989) observed three species of parasitoids, seven predators and three pathogenic fungi on this pest in India. In nature the nymphs and pupae are parasitized by *Prospaltella smithi* Silvestri, *Pleropteryx bemisiae* Mani and *Eretmocerus masii* Silvestri; besides these are also preyed upon by *Brumus* and *Chrysopa* species. Early stages coccinellids, *Menochilus* sp. And *Verania* sp. and late stage predators like phytoseiid mites have been recorded.

Prophylactic use of insecticides to suppress the population build up of whiteflies has been recommended with dimethoate

30 EC @ 250 ml or oxydemeton methyl 25 EC @ 300 ml/acre. Ethion 50 EC @ 800 ml or triazophos 40 EC @ 600 ml/acre have been proved effective in controlling whitefly *B.tabaci* in cotton (Singh *et al.*, 1994).

THRIPS

Various species of thrips viz., *Ayyaria chaetophora* Karny, *Caliothrips phaseoli* (Hood), *Taeniothrips distalis* (Karny), *Megaleurothrips usitatus* Bagnall, and *Frankliniella schultzei* (Trybom) = (*dampfi* Priesner) have been reported on guar crop.

Thrips are almost completely cosmopolitan in distribution. The damage caused by thrips is usually not serious but can be severe and sometime results in serious losses.

Leaf feeding thrips

Ayyaria chaetophora Karny has been reported damaging clusterbean leaves (Ananthakrishnan, 1971). It is a polyphagous species and feeds on a number of other cultivated and wild host plants. It completes its life cycle in 15 to 20 days. Egg, nymphal, pre-pupal and pupal period range between 4-5, 7-10, one and 3-4 days, respectively (Ananthakrishnan, 1971).

Damage is caused by rasping the leaf tissues and sucking the sap as its exudes. The surfaces of other leaves where many thrips have fed, develop a silvery sheen, largely due to air occupying the emptied cell cavities. The sillvering is emphasized by the lens like effect of epidermis above the shattered cells. The discoloured area coalesce and the leaves may dry and fall prematurely.

Bean thrip

Caliothrips phaseoli (Hood) (=*Heliothrips phaseola)* is another heliothripine polyphagous species reported infesting clusterbeans. The nature and symptom of damage are similar to these caused by other leaf feeding thrips.

The eggs are laid singly inside the tender tissues of the plant, mostly on the leaves with their saw like ovipositor. A single female may lay 25-50 eggs in its life time. Depending on the temperature, the incubation period ranges between 3-16 days. There are two larval instars, a pre-pupal and a pupal stage in the development. The total life cycle, however, ranges from 11 to 14 days from first instar to adult. There are several generations in a year.

The Blossom Thrips viz., *Frankliniella schultzei* (Trybom) (= *F.dampfi*), *Megaleurothrips* (=*Taeniothrips*) *distalis* (Karny) and *M.usitatus* (Bagnal) have been reported to cause damage to clusterbeans.

Management

Plants under water stress are susceptible to thrips attack. Hence, adequate irrigation in the clusterbean crop during hot season can reduce the losses.

Careful adjustment in time of planting and harvesting can diminish the severity of attack. Early planting of the crop usually withstands the attack because the crop becomes well established by the time thrips attack the same. However, the benefit of early sowing would depend upon the climate, soil and variety of the crop. Wild grasses provide a permanent and uncontrollable source of thrips, hence should be destroyed.

Often large proportions of field population of thrips are killed by predators and parasites. Chrysopid larvae are among the most voracious predators of thrips. Mirid bugs also attack thrips more often in warmer regions. To control the thrips, if and when necessary, spraying with 0.3% dimethoate, phosphamidon, formothion, thiometon or endosulfan has been recommended (Jotwani and Bhutani, 1977). Spraying of 250 ml malathion 50 EC in 250 litres of water per acre has also been recommended to control the pest as soon as it appears.

LEAF PERFORATOR *Dichomeris ianthus* Meyr.

This pest has established itself as a pest of clusterbean in the states of U.P., Punjab (Dhaliwal 1980) and Haryana (Singh *et al.*, 1981). The damage caused by this insect-pest has been observed to be more severe during early stages of the crop growth (Singh *et al.*, 1982).

The newly emerged larva crawls to some distance and settles down either near the midrib or a thick vein of the leaf where it spins a web. Under the cover of the web, it starts nibbling the leaf epidermis which results in minute holes. It is because of this peculiar mode of damage, the insect is called leaf perforator. The grown up larvae may web in the central whorl of the plant or two or more leaves on which they feed from inside.

Life cycle

The adults are straw coloured with two deep brown spots on each forewing with a wing expanse of 13.2 x 12.0 mm. The female moth lays creamy white, barrel shaped eggs marked with broken longitudinal ridges. Eggs are laid either singly or in clusters (2-8) along the mid rib or veins on both sides of the leaves and other tender parts of the plant. A single female may lay upto 175 eggs in her 13 days life. The incubation period lasts for 3-4 days. The early instars are yellow in colour with black head whereas the fifth instar larvae turn green with pinkish spots on 5th and 6th abdominal segment. The total larval period varies between 13-22 days depending upon the environmental conditions. The pre-pupal stage is light brown in colour and lasts for one day only. After the prepupal stage, a silken cocoon is fabricated in the folded leaves where pupation takes place. Pupal stage is completed in 6-8 days. The pest completes its life cycle in 29 days. The pest can have three generations in one cropping season on clusterbean (Ram & Verma, 1991).

The pest can be controlled by foliar application of Carbaryl (0.2%). Dusting the crop with Malathion 5 D @ 25 kg a.i./ha checks the incidence of the pest to some extent.

BOLL WORMS

Helicoverpa (=Heliothis) armigera Hubner and

Heliothis zea Buddie

Helicoverpa spp. are polyphagous insects damaging more than 60 cultivated plant species. Though this pest has been reported to damage guar crop in India (Singh *et al.,* 1981), yet there is no evidence of its causing any economic losses to this crop. In India, the abudance and peak activity of *H.armigera* is recorded during March-April, July-August, December-January and end of February. Guar has also fallen within the ambit of its host range during *Kharif* season.

Damage

The larvae are polyphagous and feed first on leaves and fruiting bodies. Later on, the larvae bore into the pod locules and damage it completely. A single larva may destroy 30-40 pods before it reaches maturity.

Life cycle

The eggs are yellow white; laid singly on tender plant parts. A single female may lay as many as 730 eggs. The incubation period varies from 2-6 days. The neonate larvae are translucent and yellowish white in colour with faint yellowish longitudinal lines. The full grown larva is about 35-42 mm long, pale green in colour with one broken stripe along each lateral side of the body with another line on the dorsal side. The general colour is variable, ranging from green to reddish brown or even black. There are normally six larval instars which take about 11-15 days to complete their development. The mature larva burrows into the soil to varying depths (2.5-17.5 cms) depending upon the type of soil for pupation. In dry heavy soils, the larvae can not bore and pupate in soil cracks. The pupal period lasts for 8-15 days in an active season which may prolong in winter. There may be as many as 8 generations from May-November.

Management

No study seems to have been made for the management of *H. armigera* especially in guar crop. However, in other field crops, the manipulation in cultural practices, host plant resistance or biological control of this pest have been standardised.

Endosulfan (0.07%), fenvalerate (0.02%), cypermethrin (0.01%), deltamethrin (0.004%) or neem seed kernel extract (5%) have been found effective (Lal *et al.,* 1985; Reed *et al.,* 1987). The ovicidal action of quinalphos (0.05%), entrofolan (0.1%) or triazophos (0.1%) has practical value in *H.armigera* management.

H.armigera larvae also suffer from nuclear polyhedrosis virus (NPV). Hence, use of HNPV at 250-500 larval equivalents (Les')/ha containing 0.5% Jaggery and 0.1% sticker (Ranipal) at weekly interval during peak oviposition period has been found effective against this pest (Reddy and Krishnamurthy, 1989).

In India a larval parasite, *Campolestis chloridae* Uchida has been reported to be a major natural mortality factor of *H.armigera* in leguminous crop.

JASSIDS/HOPPERS

E. motti (= *kerri*) is the most prevalent species of leaf hopper in India. It is polyphagous pest and has been recorded from numerous plant species in India (Mann & Sohi, 1993). The pest appears on guar during the rainy season and attacks throughout the crop growth.

Damage

Both nymphs and adults suck the sap from leaves and tender parts of the plant resulting in yellow discoloration of leaf edges followed by cupping of leaves. In case of heavy infestation, the leaves roll down at the edges, dry and fall down. Besides devitalising the plants and affecting the yield adversely, it also acts as a vector of plant viral diseases.

Life cycle

Adults are 3 mm long and yellowish green in colour. A female lays 25 to 60 eggs in 25 to 30 days within the leaf veins on the upper surface of the leaf which hatch in 3-11 days. Nymphal development is completed in an average of 25 days. Females live longer (132 weeks) than males (2 weeks). Life cycle is completed in 15 to 45 days and there may be 10 to 12 overlapping generations in a year. Intermittent rains, cloudy weather and high temperature favour the multiplication of this pest. The damage caused to *kharif* season crops is, thus, more pronounced than *rabi* or spring/summer crop (Chaudhary *et al.*, 1981).

Management

Prohylactic spraying with any one of the systemic insecticides viz., 0.05% dimethoate or formothion, 0.03% phosphamidon or oxydemeton-methyl check the jassid activity effectively. If need be, spraying with one of the above insecticides may be repeated at an internal of 10 to 14 days to obtain complete check of the pest.

In nature, *Chrysopa cymbela,* black ants, *Componotus* sp. and a spider mite, *Distina albida* have been observed predating upon nymphs and adults of *E.kerri*.

WHITE GRUBS

Holotrichia consanguinea Blanch

White grubs, commonly known as beetle grubs or chafer beetles or June beetles are among the very destructive polyphagous pests. *H.consanguinea* is predominant species particularly in the north western state of Haryana, Rajasthan, Punjab and H.P. (Brar and Sandu, 1981). Its host range includes groundnut, sugarcane, millets, maize, chillies, okra, brinjal and guar.

Damage

Both the grubs and adult beetle damage the crop. The grubs feed on roots, root nodules and fine rootlets of the plant and may also girdle the main root. The adults hide in the soil during the day time and come out during night to feed on the foliage of field crops and trees such as *ber, neem* etc.

Life cycle

The female lays eggs in soil singly upto a depth of 10 cms during February and March which hatch within 7-10 days. Newly emerged grubs are shining white in colour with black head. Larval development is completed in about 56-70 days. After the monsoon, the full grown larvae migrate below the soil surface from 30-150 cms deep and form earthen cells to pupate. The pupa is creamy white and semi-circular. Pupal stage lasts for 15 days. The adults emerging in November do not come out of the soil till next June and become abundant in November during for the coming rainy season. There is only one generation in a year.

Management

. Campaign for manual collection and destruction of photopositive beetles at night after the first shower of rains in June-July is the most effective method for the control of beetles (Amin, 1988). Deep ploughing of fields exposes pupae to predatory birds (Labana, 1987)

Anthia sexguttata Fab. predates upon the adult beetles (Yadava, 1981); the Indian toad, *Bufo melanostichus* Schn. and the gecko, *Gecko gecko* (L.) prey upon the adults. Birds viz., The crow, *Corvus splendens* Vieil and Indian Maynah *Acridotheres tristis* (L.) are predators of exposed grubs at the time of ploughing. One fungus, *Metarrhizum anisopliae* (Metsch.) and a bacterium *Bacillus popillae* Dutky have been found pathogenic on the adults and grubs, respectively.

Spraying of host trees and crops with Carbaryl (0.1%) monocrotophos (0.05%) or quinalphos (0.025%) gives effective control of adults. Seed treatment at the time of sowing with chlorophyriphos 20 EC and isofenphos 40 SD at 2.5 g a.i./kg seed has been recommended for the control of grubs. Application of chlorphyriphos 20 EC @ 4 lts/ha or quinalphos 25 EC @ 4L/ha with irrigation water also controls the grubs in standing crop (Gupta and Yadava, 1987).

STEM WEEVILS

Several species of stem weevils viz., *Alcidodes bubo* (Fab.), *A.fabricii* (Fab.), *A. collaris* (Pasc.), *A. pictus* (Bohem.), *Apion leopardus* (Oliv.), *Blosyrus inequalis* (Bohm.), *Cyrtozema cognata* (Marsh.), *C.dispar* (Marsh.), *Myllocerus maculosus* (Desb.) have been recorded on clusterbean. Of these, *A. bubo* and *M.maculosus* are comparatively more abundant.

Damage

Both adult and grub damage the crop making holes in the leaves and numerous punctures on the stem thereby weakening the plant considerably. The grubs often prove more destructive than adults. They bore into the stem and make irregular galleries, eventrually causing uneven growth of the stem and gall like swellings at the place of injury. The infested plants present a sickly appearance and growth is stunted. When the attack takes place at the seedling stage, the whole plant succumbs. At time, pods are also infested and damaged.

Life cycle

Elongated, oval and pale yellow eggs are laid singly in the cavities formed by gnawing the soft tissues on the apical portion of tender shoots and also on guar pods. The incubution period lasts for 6-7 days. The newly emerged grubs are pale white in colour with pale brown head. The grubs pass through six instars and becomes full grown in about 25 days. Pupation takes place inside the larval galleries and pupal duration varies from 9-11 days. Pupation takes place inside the larval galleries and pupal duration varies from 9-11 days. Life cycle is completed in 49-57 days.

Management

Mechanical methods like destruction of immature stages by removing and destroying infested plants, collection and burnin of stubbles are effective in bringing down the incidence.

Prophylactic use of contact insecticides like malathion 5 D

or trichlorophon 5 D or spraying the crop with 0.2% Carbaryl help in keeping the pest population under control.

The grubs are parasitized by *Eurytoma pigra* Girault and *Metastenomyia juliani* Girault.

Cotton Grey weevil *M.maculosus* a polyphagous pest (cotton, sorghum, pearl millet, maize etc.) has also been reported to occur regularly on clusterbeans (Butani, 1980).

Damage

The grubs remain in soil and feed on roots. Adults feed on leaf margins; sometimes damage the flower buds completely. Infestation at early stage of plant growth results in the seedling mortality.

Life cycle

Adults are 8-9 mm long, pale green in colour, scaled bodies; ultra with variable dark brownish-grey patches and lines appear punctuated in rows. Ovoid and light yellow eggs are laid 80-100 mm deep in soil. The grubs are stout, fleshy, 7 to 8 mm long and hold their bodies in more or less of a curved position. Incubation period is 3 to 11 days in summer and 8 days in winter. The larval period lasts for 28 to 34 days. Pupal period takes about 5 to 7 days. Total life cycle is completed in 6 to 8 weeks and there are 5 generations in a year. The pest passes winter in grub stage 40 to 50 cms below the soil surface.

Management

Generally no special control measures are adopted to control this pest. To prevent population build up and checking the carry over of the pest, regular interulture of the crop (70-80 cms deep), to expose the grubs, has been suggested. In nature, grubs are parasitized by *Dinocampus mylloceri* Wilkinson.

SUMMARY

In tropics, Cluster bean is among the most important crops raised under sustainable agriculture system. Management of insect pests is essential for successful cultivation of this crop, since 41 genera from 16 families and eight orders have been reported to be associated with this crop world over. Among the foliage feeders, hairy caterpillars, thrips, leaf perforators, bollworms etc. are important. Sucking pests viz., jassids, aphids, whiteflies not only reduce the seedling vigour by sucking the cell sap but also transmit several viral diseases. Normal photosynthesis is affected due to sooty mould which develop on honeydew excreted by whiteflies and aphids. Bollworms damage the fruiting bodies and later bore into pod locules to reduce the yields significantly. White grubs are especially serious after first showers during monsoon. Information on host plants, damage, life cycle and management (cultural, mechanical, biological and chemical) of these insect pests has been reviewed.

REFERENCES

Abbas, S.R. 1967. A special morphological structure "Cutting organ" in the pupa of *Dyodiplosis glomeratii* Grover (Diptera; Cecidomyiidae). *Indian J. Ent.* 29: 99-101.

Abbas, M. and Afzal, M. 1946. Cotton jassid (*Empoasca devastans* Dist.) in the Punjab VI. Species found on the cotton plant in the Punjab. *Indian J. Agric. Sci.* 15: 119-124.

Ahlawat, D.S., Saini, R.K. and Singh, H. 1993. Biology of the red hairy caterpillar *Amsacta moorei* Butler (Arctidae: Lepidoptera). *Ann. Biol.* 9: 118-122.

Amin, P.W. 1988. Insect and mite pests and their control. In. *Groundnut* (ed. Reddy, P.S.) ICAR, New Delhi, pp. 393-452.

Ananthakrishnan, T.N. 1971. Thrips (Thysanoptera) in agriculture, horticulture and forestry – Diagnosis, bionomics and control. *J. Sci. Ind. Res.* 30: 113-146.

Arora, G.L. 1977. Taxonomy of the Bruchidae (Coleoptera) of northwest India. Part I, Adults. *Orient. Insects* Supl. 7: 32 pp.

Atwal, A.S. and Dhaliwal, G.S. 1997. *Agricultural pests of South Asia and their management*, Kalyani Publishers, Ludhiana, 487 pp.

Bakhetia, D.R.C. and Sidhu, A.S. 1977. Biology and seasonal activity of groundnut aphid, *Aphis craccivora* Koch. *J. Res. PAU* 14(3): 299-303.

Brar, K.S. and Sandhu, G.S. 1981. Note on the effect of dates of sowing on the damage by *Holotrichia consanguinea* Blanchard to groundnut. *Indian J. agric. Sci.* 51: 466.

Butani, D.K. 1980. Insect pests of vegetables and their control – Clusterbean. *Pesticides*, 14(10): 33-35.

Butani, D.K. and Jotwani, M.G. 1984. *Insects in Vegetables*. Periodical Expert Book Agency, Delhi 356 pp.

Buti, S., Jabbar, A. and Ahmed, A. 1982. Biology and ecology of Empoasca kerri (Typhlocybinae: Cicadellidae) a pest of guar, *Cyamopsis tetragonoloba* in Pakistan. *Biologia* (Lahore) 27(2): 257-264.

Chari, M.S. 1971. Nature and extent of damage caused by the larvae of clusterbean pod gall midge *Asphondylia cyamopsii* n sp. On different varieties of clusterbean. *Cyamopsis tetragonoloba* (L.) Taub. *Madras agric. J.* 58: 892-893.

Chaudhary, K.P., Singh, A.K. and Yadava, L.S. 1981. Effect of disulfotan granules on pest complex, crop growth and yield of green gram. *Indian J. Ent.* 43: 369-372.

Chhabra, K.S. and Kooner, B.S. 1989. Sources of whitefly *Bemisia tabaci* G. and yellow mosaic virus resistance in mung bean. *Vigna radiata* W. *Trop. Grain Legume Bull.* 19: 26-29.

Chhabra, K.S., Lal, S., Kooner, B.S. and Verma, M.M. 1993. *Insect pests of pulses – identification and control manual.* PAU Ludhiana & Directorate of Pulses Res. Kanpur 88 pp.

COPR, 1981. *Pest control in Tropical grain legumes.* Centre for Overseas Pest Control, London. 127-206.

Dennis, R.E. 1976. Growing guar in Arizona. Univ. of Arizona, Dept. Plant Sci. *Agril. File* FC 288.0.

Dhaliwal, J.S. 1980. Important insect-pests of guar in the Punjab State. *Guar Newsl.* 1: 18.

Gill, R.J. 1990. The morphology of whiteflies, 13-16. In *Whiteflies: Their bionomics, pests and management* (ed. Gerling, D.), Intercept, UK 348 pp.

Gupta, B.M. and Yadava, C.P.S. 1987. Chemical control of white grub, *Holotrichia consanguinea* Blanchard infesting groundnut. *Indian J. agric. Sci.* 57: 742-744.

Hussain, M.A. and Khan, A.W. 1945. The citrus Aleyrodidae (Homoptera) in the Punjab and their control. *Mem. Entomol. Soc. India.* 41 pp.

Jotwani, M.G. and Butani, D.K. 1977. Insect pests of leguminous vegetables and their control. *Pesticides* 11(10): 35-38.

Kaiser, W.J. 1979. Natural infection of cowpea and mung bean by alpha-alpha mosaic virus in Iran. *Pl. Dis. Rep.* 63(5): 414-418.

Khurana, A.D. and Kaushik, H.D. 1991. Bioefficacy of insecticides against *Aphis craccivora* Koch. and *Agrotis ipsilon* (Hufn.) on chickpea. *J. Insect Sci.* 4: 193-194.

Krishnakumar, N.K. and Sardana, H.R. 1988. Pest management in solanaceous vegetables. In: *Annu. Report*, 1988 IIHR, Bangalore, pp. 64.

Kumar, A. and Nigam, P.M. 1989. *Economic and Applied Entomology*, 1st Edn. Emkay Publications, Delhi pp. 201.

Labana, K.S. 1987. It is time to control the white grub. *Indian Fmg.* 37(4): 31.

Lal, S.S., Yadava, C.P. and Dias, C.A.R. 1989. Effect of planting density and chickpea cultivars on the infestation of the black aphid, *Aphis craccivora* Koch. *Madras agric. J.* 76: 461-462.

Mann, J.S. and Sohi, A.S. 1993. New records of Asian Empoascini (Homoptera: Typhlocybinae). *J. Insect Sci.* 6: 232-238.

Panwar, V.P.S. 1995. Agricultural Insect Pests of Crops and their Control. Kalyani Publishers, New Delhi 286, 286 pp.

Patel, H.K. and Chari, M.S. 1970. New record of gall midge Asphondylia sp. As a serious pest of clusterbean *Cyamopsis tetragonoloba* (L.) Taub. *Indian J. Entomol.* 32: 90-91.

Patel, N.C., Valand, V.M. and Patel, S.N. 1995. Ovicidal action of some new insecticides against eggs of *Helicoverpa armigera* (Hubner). *Indian J. Pl. Prot.* 23: 99-100.

Prasad, D., Singh, K.M., Katiyar, R.N. and Singh, R.N. 1988. Impact of intercropping on plant growth, pest incidence and crop yield of pea, *Pisum sativum* Linn. *Indian J. Ent.* 49: 153-172.

Pruszynski, S. and Mrowezynski, M. 1990. Trial result on the use of Lambda-cyhalothrin for pest control in agricultural crops in Poland. In *Proc. Brighton Crop Prot. Conf., Pests and Diseases* Vol. II BCPC Publication pp. 715-720.

Rai, B.K., 1976. *Pests of Oilseed Crops and Their Control*. ICAR, New Delhi.

Ram, P. and Verma, A.N. 1991. Biology of *Dichomeris ianthus* Meyr – a pest of clusterbean (*Cyamopsis tetragonoloba* (L.) Taub.) in Haryana. *Agric. Sci. Digest.* 11(4): 181-184.

Rao, N.V., Reddy, A.S. and Reddy, D.D.R. 1990. Relative toxicity of some insecticides to cotton whitefly. Bemisia tabaci, Indian J. Pl. Prot. 18(1): 97-100.

Rao, R. 1917. Clusterbean (*Cyamopsis psoralioides*) (Guar-Hind). *Proc. II Entomol. Meet*, Pusa Calcutta, India, p. 60-61.

Reddy, G.P.V. and Krishnamurthy, M.M. 1989. Insect Pest Management in Cotton. *Pesticides*, 23(10): 18-19.

Reed, W., Cardona, C., Sithanantham, S. and Lateef, S.S. 1987. Insect pests and their control. In *The Chickpea* (eds.) Saxena, M.C. and Singh, K.B., CAB Int., Wallingford, UK.

Rogers, C.E. 1972. Midges associated with Guar in Texas and Oklahoma. *Ann. Rnyomol. Dov. Smrt.* 65: 1203-1208.

Saini, R.K. 1993. Development and survival of red hairy caterpillar, *Amsacta moorei* Butler on some cultivated plants and weeds in Haryana. *J. Insect. Sci.* 6: 64-68.

Saini, R.K. 1995. Evaluationof some promising insecticides against red hairy caterpillar, *Amsacta moorei* Butler. *J. Insect. Sci.* 8: 220-222.

Saini, R.K. and Verma, A.N. 1991. Effectiveness of light traps in suppressing populations of red hairy caterpillar *Amsacta moorei* Butler. *Haryana agric. Univ. J. Res.* 21: 250-252.

Seikh, A.G., Rather, A.J. and Mattoo, M.S. 1990. Evaluation of various insecticides for the management of cowpea aphid, *Aphis craccivora* (Koch) infesting moong. *J. Aphidology* 4(1 & 2): 36-38.

Singh, H. and Yadava, T.P. 1985. Strategies for the management of insect-pests in sesame to increase its production. In *Oilseeds Production, Constraints and Opportunities* (eds.) Srivastava, H.K., Bhaskaran, S., Vatsya, B. and Menon, K.K.G. Oxford & IBH Publ. New Delhi.

Singh, S.P., Singh, J.V. and Singh, V.P. 1996. Screening of clusterbean genotypes for resistance to whitefly, *Bemisia tabaci Genn. Forage Res.* 22: 59-62.

Singh, S.P., Bhanot, J.P. and Saini, M.L. 1982. New record of leaf perforator, *Dichomeris ianthus* Meyr. (Lepidoptera: Gelechidae) on guar from Haryana. *Guar Newsl.* 3:11.

Singh, S.P., Verma, A.N. and Bhanot, J.P. 1981. Insect and mite pests of guar and their control. *Guar Newsl.* 2: 47-51.

Srinivasan, K. 1994. Recent trends in insect-pest management in vegetable crops. 4: 34-372. In *Trends in Agricultural Insect pest management* (eds.) Dhaliwal, G.S. and Arora, R. Commonwealth Publishers, New Delhi.

Srivastava, A.S. and Goel, B.K. 1962. Bionomics and control of red hairy caterpillar (*Amsacta moorei* Butl.) *Proc. Nat. Acad Sci. India* 32B: 97-100.

Streets, R.B. 1948. Diseases of guar (*Cyamopsis psoralioides*). Phytopathol. 38:198.

Subramanian, T.R. 1959. The biology of the weevil *Alcidodes bubo* (Febricius) (Coleoptera: Curculionidae). *J. Bombay Nat. Hist. Soc.* 54: 581-626.

Thotappilly, G. and Rossel, H.W. 1985. Worldwide occurrence and distribution of virus diseases. 155-171. In. *Cowpea Research, Production and Utilization* (eds.) Singh, S.R. and Rachie, K.O. John Wiley, London, 460 pp.

Whistler, R.L. and Hymolvitz, T. 1979. *Guar: Agronomy, Production, Industrial use, and Nutrition.* Purdue Univ. Press, Indiana, USA.

Yadava, C.P.S. 1981. *Integrated Control of White grubs.* Final Res. Report. ICAR Project. Dept. Entomol. Univ. of Udaipur, Jobner.

Yadava, C.P.S., Bhatnagar, K.N. and Srivastava, B.P. 1966. Incidence of *Amsacta moorei* Butler in the semi-arid zones of Rajasthan. *Indian J. Ent.* 28: 274.

CHAPTER - 10

CHEMISTRY AND NUTRITIVE VALUE OF GUAR

U.N. Joshi

Clusterbean (*Cyamopsis tetragonoloba* (L.) Taub), a drought hardy leguminous crop, has emerged as an industrial crop, which helps in earning foreign exchange for India. For instance, guar gum export reached its peak of 101.09 thousand ton providing foreign exchange worth Rs 543.80 crore during 1997-98 (Table 1) which is the testimony of its importance as foreign exchange earner. Traditionally, its green pods are used as vegetable for human consumption, plant as fodder for livestock (Joshi *et al.*, 1991), seeds for the extraction of gum, which serves as a useful source of hydrocolloids for various uses and guar meal as feed for non-ruminants. Being a leguminous crop it is also used for soil improvement. Empty guar pods are better in forage quality than guar fodder (Arora *et al.*, 1980).

GUAR SEED

Guar seed consists of seed coat (14-17%), endosperm (35-42%) and germ (43-47%). Guar seed has attained an important place in industry because of its galactomannan rich endosperm. The hull and germ portion of guar seed are termed as guar meal. An average composition of seed components of guar has been reported by Das and Arora, 1978 (Table 2). Various workers have also reported variability in the chemical composition of guar seed (Mehrotra *et al.*, 1975). Mehta and Ramakrishnan (1957) reported 4.2% oil in seeds, whereas the maximum value of oil content (9.8%) in guar seed cotyledons has been reported by Shinde and Bhargava (1968). Joshi *et al.* (1990) also observed considerable variability in oil content (3.06 to 7.5%) in seeds of various species of guar.

Table 1. Export of guar gum

Year	Quantity of 000 tonnes	Values (Rs. in crores)
1985-86	46.58	36.19
1986-87	28.34	33.67
1987-88	43.95	80.21
1988-89	40.53	N.A.
1989-90	N.A.	N.A.
1990-91	N.A.	N.A.
1991-92	N.A.	N.A.
1992-93	62.02	72.72
1993-94	74.42	140.67
1994-95	67.33	141.99
1995-96	82.94	226.36
1996-97	94.63	354.11
1997-98	101.09	543.80

Source : APEDA, 1999, New Delhi

Table 2. Composition of the components of guar seed

Seed part	Protein % (N x 6.25)	Ether extract %	Ash %	Moisture %	Crude fibre %	Type of carbohydrate
Hull (14-17%)	5	0.3	4	10	35.0	D-Glucose
Endosperm (35-42%)	5	0.6	0.6	10	1.5	Galactomannan
Germ (43-47%)	55.3	5.2	4.6	10	18.0	D-Glucose

Source : Das and Arora (1978).

GUAR GUM

The endosperm which is rich in galactomannan type of gum got commercial boost during world war II, when the supply of locust bean gum, the traditional raw material was adversely affected and the best possible substitute was found as guar gum. Chemistry and industrial utilization of guar gum have been reviewed by Das and Arora (1978) and Joshi and Arora (1993). The guar gum consists of considerable fractions of Zn and Cu metals and protein to the tune of almost 5.0% (Table 3).

Table 3. The typical maximum composition of high grade guar gum

	Per cent	Metals (ppm)	
Water	12.0	As	0.20
Protein	5.0	Pb	0.03*
AIR	3.0	Cu	2.50
Ash	0.6	Zn	5.60
pH (1% solution)	6.5	Cd	0.05*
Viscosity	3000-5000 mPa s		
(1% solution fully hydrated, RVT Brookfield, 20 rpm, 25°)			

Source : Joshi and Arora (1993).
*Calculated value
AIR = Acid Insoluble Residue.

EXTRACTION AND PURIFICATION OF GUAR GUM

A number of techniques were proposed for extraction and purification of guar gum. The most reliable and accurate technique involves is the extraction and purification of galactomannan which is thereafter alcohol precipitated, dried and weighed (Sikka and Johari, 1969). Patel (1958) obtained patent for guar gum isolation and isolated it by loosening and removing husk of seeds using chemical agent and also by cooking the seeds in water. In an Indian patent, the process to obtain refined guar gum in pan-mills without using chemicals has been described (Jiwaji Industrial Research Laboratory, Gwalior, 1960). The production of guar endosperm and guar meal from seeds is shown diagrammatically in Fig. 1. The round whole seeds are fed into a mill having two grinding surfaces travelling at different speeds. The seeds are split into two halves i.e. endosperm with adhering hull and the germ which is very friable, shatters and is shifted through, as fine material. The crude crack (endosperm + hull) is heated to soften the shell and is then fed into hammer mill which can either shatter the hull away from endosperm or into a mill, which can abrade the hull away from the endosperm. Any remaining germ is also pulverized and the resultant split is essentially pure endosperm. The hull of the guar seed, thus, can be removed quite easily from endosperm halves (Joshi and Arora, 1993).

Purification of the gum is generally done by repeated precipitation with alcohol. The isopropyl alcohol is mostly used in industrial processes. Kendurkar (1970) refined guar gum by adding (200 g) gum to boiling solution of ammonium oxalate

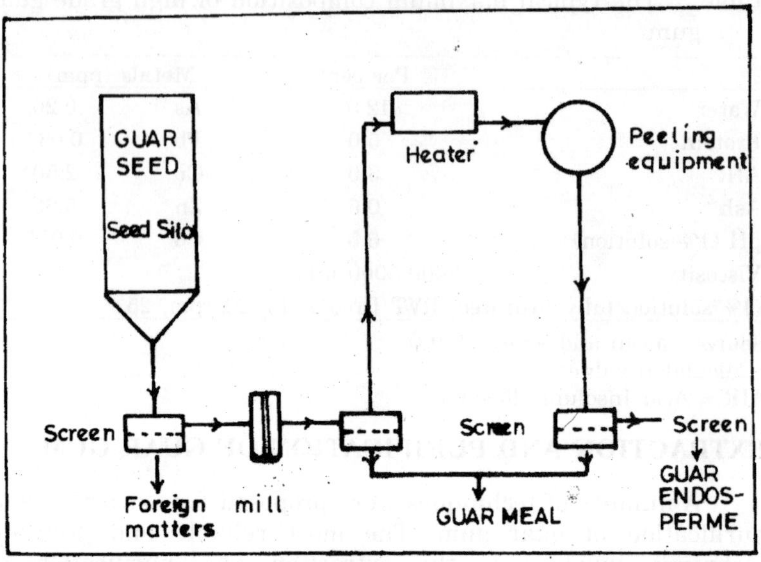

Fig. 1. Flow diagram for production of guar endosperm and guar meal from the guar seed

(0.2%, 100 ml). The mixture boiled for 2 minutes stirred for 5 min and mixed with 500 ml of boiling water and again stirred for 5 minutes. The separated husks were removed by washing with water. The blue coloured gum was kept in 200 ml of aqueous solution of SO_2 (0.3%) for 10 minutes and the solution drained. The gum was dried at 60° for one hour to obtain the refined product. Das and Arora (1978) reviewed that gum can be purified by mixing 1 kg of crude gum with 10 kg of 1:2 (vol.) α-picoline-CCl_4 mixture and drying the filtrate at >50°C under reduced pressure to give 30 g of purified gum. Relan *et al.* (1991) purified the guar gum by eighty per cent alcohol by repeated washings at low temperature using soxhlet apparatus.

Estimation of guar gum

The methods for estimation of guar gum on a laboratory scale were developed by Sikka and Johari (1969) and Mittal *et al.* (1971). Das *et al.* (1977) extracted the gum by autoclaving the guar flour for one hour with mercuric chloride solution and estimated the soluble sugars by spectrophotometer. The technique for gum estimation in guar seed by near infra red reflectance spectroscopy was also reported.

Composition and structure

The endosperm separated from the hull and germ, is ground to fine particle size and marketed as guar gum and called as "guaran". The crude gum is greyish white powder and may contain small amount of proteinaceous matter. The major component of guaran is galactomannan (78-82%), protein (4-5%), crude fibre (1.5-2.0%), ash (0.5-0.9%), fat (0.5-0.75%) and moisture (10-13%) (Das and Arora, 1978, Arora, 1989). The galactomannan gum or guaran contains 34.6 per cent D galactopyranosyl units and 63.4 per cent D-mannopyranosyl units and consists of a linear chain of D-mannose units linked together by ß (1-4) glycosidic linkages. The galactose side stubs are attached as single side chains to the mannan back bone by α (1-6) glycosidic linkage. On an average, alternate D-mannose units bear a galactose unit. Both mannose and galactose have hydroxyl groups in cis-position which is responsible for cross linking property of guar gum (Fig. 2). The galactose to mannose ratio in guaran is 1:2 and information about the structure is deduced from enzymatic studies. The biodegradative hydrolysis of a galactomannan requires the presence of at least three enzymes in the germinating seeds i.e. α-D-galactosidase for removal of the (1-6) α-D-galactoside chains, ß-D-mannanase for fission of the (1-4) ß-D-mannan backbone into oligosacch- arides and ß-D-mannosidase for the complete hydrolysis of D-manno-oligosaccharides to mannose. These enzymes have been reported in guar seed (Sehgal *et al.*, 1973). Molecular weight of guar galactomannan determined using light scattering method by Deb and Mukherjee (1963) was 1.7×10^6. Sundarajan and Rao (1970) reported that guar gum polymer in aqueous solution has the mannan backbone in a fully extended rod conformation, whereas, Relan *et al.* (1991) reported disk like configuration of guar gum molecule at low temperature and flexible linear molecule at higher temperature. The composition (Table 3 and 4) and structure of guar gum (Fig. 2) was discussed and reviewed by Das and Arora (1978), Arora (1989) and Joshi and Arora (1993).

Properties

Chemical and physical properties of guar gum were discussed by Arora (1989) and Joshi and Arora (1993). The main properties are discussed.

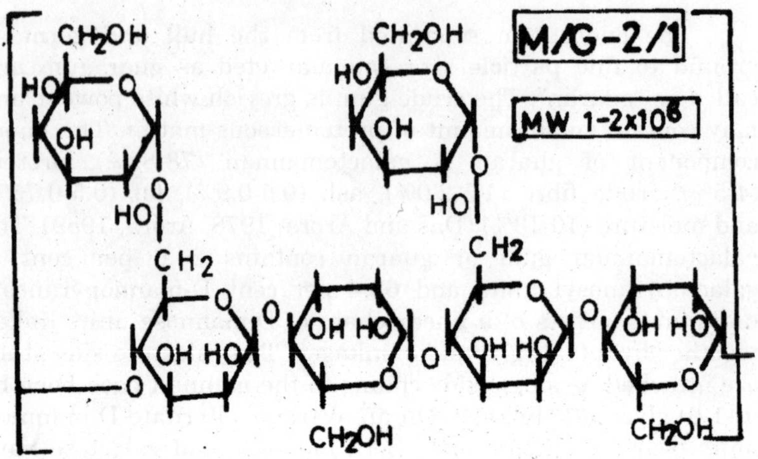

Fig. 2 Structure of galactomannan gum from *Cyamopsis* sp. (guar)

Table 4. The average amino acid composition (residues per 1000 residues) of the proteinaceous part of guar gum.

% N	0.57		
Alanine	83	Leucine	62
Arginine	50	Lysine	52
Aspartic acid	100	Methionine	8
Cystine	3	Phenylalanine	30
Glutamic acid	126	Proline	46
Glycine	142	Serine	82
Histidine	26	Threonine	53
Hydroxyproline	14	Tyrosine	28
Isoleucine	57	Valine	58

Source : Joshi and Arora (1993).

Viscosity

Guar gum hydrates very rapidly in cold water system to give highly viscous solutions, because the side groups hold the main mannan chain far apart to prevent noncovalent interactions. The viscosity of guar gum solution varys with shear rate, which is typical of non-newtonian fluids. Viscosity of 0.3 per cent solution changes only slightly, whereas solutions of

1 per cent concentration or higher changes markedly with change in shear rate. Viscosity depends on time, temperature, concentration, pH, ionic strength and type of agitation. The viscosity of 1 per cent solution ranged from 3000 to 5500, at 25°C. The increase in shear rate and temperature caused a decrease in viscosity. The maximum viscosities of guar gum dispersions are achieved at tempera- tures of about 25-40°C. The viscosity increases linearly (Newtonian) with increase in concentration upto 0.5 per cent. Thereafter, guar gum solution behaves as non-Newtonian solution mainly due to complex surface attractions. For a typical solution, doubling the concentrations gives about ten fold increase in viscosity. The non-ionic nature of the molecule is responsible for the almost constant viscosity of solutions in the pH range of 1.0 to 10.5. The gum being nonionic, is not susceptible to salting out. Stability of solutions is good with the salts containing mono, di and trivalent cations and the viscosity increases with increasing concentrations of these salts. The effects of anions such as nitrates, chlorides and sulphates on the viscosity of guar solutions are similar i.e. the viscosity increases with anion concentration. Many polyvalent cations such as calcium, aluminium and chromium insolubilize the guar gum in solution at certain pH levels and may form a gel. The stability of solutions or dispersions are improved by using metal ion sequestering agents, particularly, those capable of inactivating iron ions. Citric, tartaric and orthophosphoric acids are preferred additives in an amount of 0.25 to 0.5 per cent based on weight of gum. Salt and sugars are the two most widely used ingredients in foods and both compete for water and decreased the rate of dissolution of guar gum (Das and Arora, 1978, Arora, 1989).

Hydration rate

The rate of hydration of guar gum varies. For achieving maximum viscosity of solution of guar gum, a hydration time of 2 hours is required. However, if quick initial viscosity is needed then very fine mesh guar gums are used but still considerable time is needed for maximum hydration and viscosity. Gupta and Bhatia (1970) studied the water vapour sorption and desorption on the gum and reported that hysteresis was due to the blockage effect of small capillaries.

Other properties

Its (guar gum solution) specific colloidal nature gives the solution an excellent thickening power which is 6 to 10 times higher than that obtained from starch. It is stable over a wide range of pH. It also improves flowability and pumpability of fluid. It is a superior frictionless reducing agent (APEDA, 1999).

Uses of Guar Gum

Guar gum has vast uses in various ways as in explosives, petroleum industry, textiles, paper, coal mining, fire fighting, pollution control, agriculture etc. The detailed uses have been briefed below:

APPLICATION OF GUAR GUM IN DIFFERENT INDUSTRIES

Product	Form of gum	Functions
1. Explosives		
Explosive	Crosslinked guar gum	Water remover
Semidynamite blasting explosive	Guar gum	Plasticity improver
Water resistant explosive	Hydratable guar gum and self complexing guar gum	Gelling agent
Free flowing granular explosive	Guar gum	Thickener
Nitroalkane explosive	Nitrate ester of guar gum	Thickening agent
Blasting agent	Guar gum with cross linking agent	Thickening agent
Slurry explosive	Guar gum or its hydroxy-propyl derivative	Thickening agent
Gel explosive	Guar gum	Gelling agent
Extrudable primer	Guar gum	Increases viscosity
Explosive mixture	Cyanoethyl ether of guar gum	Thickening agent
Blasting slurry	Guar gum with inorganic oxidizer salts	Thickening agent
Blasting agent	Guar gum with transition metal ions	Increases viscosity
Explosive gel	Guar gum	Improves resistance to water and aging
2. Petroleum industry		
Polymer association complex	Sulfonated guar gum and Me_3 N^+ guar gum	Thickener
Hydraulic fracturing fluid	Alkali refined guar gum cross linked with borax	Gelling agent
Petroleum wells	Guar gum with fused potassium	—

Petroleum well blow out	Hydroxy propyl guar gum	Thickening agent
Petroleum well fracturing	Guar gum/copolymer	Thickening agent
Fracturing fluid	Guar gum and hydroxy ethyl cellulose	Gelling agent
Fracturing fluid	Cationic guar gum	Viscosity and thermally stable
Fracturing fluid	Guar gum with boric acid and magnesium-oxide	Provides viscosity stability
Fracturing well	Potassium permanganate cross linked guar gum	—
Fracturing solution	Hydroxyl aklyl ether derivatives of guar gum	Reduces friction & increases permeability
Drilling fluids	Guar gum with mineral salts and PEG	Improve water loss
Drilling fluid	Hydroxy alkyl guar gum	Good fluid loss properties
Drilling mud	Gum ether	Thickening agent
Drilling	Guar gum with friction reducers	Thickener
Well stimulator	Derivatized guar gum	Gelling agent
Rotary drilling	Guar gum with methylene blue	Indicator of fluid life
Oil bore	Guar gum with laminar silicate starch and swelling clay	For reduction of fluid loss
Oil bores	Borate cross linked guar gum	Stable super-elastic liquid with lessened temperature sensitivity for control of lost circulation in oil field drilling operation
Oil gas water bores	Guar gum xanthomonas campestris mixtures	For plugging leaks & stabilization of likages
3. Textiles		
Printing paste	Guar gum	Thickener
Printing fluid	Guar gum and xanthomonas gum	Printing paste thickener
Cambric and muslin cotton fabric printing dye	Modified guar gum	Thickener
Textile dye printer	Guar gum derivatives (polyacrylic acid, polyacrylamide, carboxymethyl guar gum & guar gum mixture grafted with acrylic acid or acrylonitrite or acrylamide)	Thickener
Fabric printing	Guar gum PEG lauryl ether	Foam composition

Polyester printing	Derivatives of guar gum	Thickener
Textile print	Guar gum with acrylamide	Stabilizer
Gum O_2 churchr $(CO^{\cdot}{}_{20})_2$ $(M/m)^{n+}$	Sulfonated guar gum derivatives	Pigment retention aid
Thickening agent	Modified guar gum	Thickener
Fire and shrink proof textiles	Guar gum with chloro-iso-cyanuric acid	—
4. Paper		
Paper board	Guar gum	Increase fines retention
Paper board	Modified guar gum	Retention aid prior to sheet forming operation and floatation aid
	Quarterinary ammonium Guar gum	Imparts dry strength
Paper making	Quaterinized guar gum	Improves dry and wet strength
Paper making	Derivatives of guar gum (mey 'proid, SFA, mey pro bend PA109, meyprofilm PA 116)	Enhance surface and sizing Improves retention for filter barrier
Paper making	Guar gum	Improves strength
Paper	Cationic guar	Retention ·&drainage providing agent
Paper and textile	Guar gum formate	Flocculent and sizing agent
Paper and textile	Sodium salt carboxy methylated guar gum	Dry strength
5. Ore refining/metal		
Iron scrap	Amino ethyl gum	In settling fine particles colloidal
Copper electro refining	guar gum	Flocculent
Electro deposition of manganese	Guar gum with polyacrylamide	Less mechanical work
Jaguar A-4-F Jaguar A25 Jaguar HP-11 Jaguar CMHP Jaguar C13	Guar gum and its derivatives	Binder
6. Coal mining		
Coal slurry stabilizer	Esterified guar gum	Stabilizer
Slurries	Guar gum	Dispersant
Coal mining	Guar gum with boric acid/borax	For shock impregnation of coal seams

7. Tobacco		
Tobacco products	Guar gum	Adhesive
Tobacco sheet	Guar gum	Reduces irritation & strengthening agent
8. Electricals and telephone		
Gel lubricant	Guar gum	Lubricant for installation of electric and telephone cable
Cable cores	Gum with isobutylene and Carboxymethyl cellulose	Electrical insulator
9. Jute		
Sizing agent	Guar gum	Sizing agent
10. Varnish and coating		
Base for varnish and coating	Guar gum derivatives	Protective colloid
11. Ceramics		
Ceramic articles	Guar gum with ceramic powder	Paste for moulded articles
12. Fire fighting		
Fire extinguishing power	Carboxy methylated guar gum	Provides storing stability
Forest fire fighting	Guar gum with glycerol and ethylene glycol	Dispersions
Air drop forest-fire control	Guar gum with decyl sulphate & ammonium phosphate solution	Provide viscosity and stability
13. Building/construction material		
Water proofing under ground construction material	Guar gum	Gelling agent
Foamed cement material	Guar gum	Foam stabilizer
Plaster	Guar gum	Thickening agent
Concrete	Guar gum	Water proofing
14. Pollution control/water purifier		
Adsorbent in textile waste water treatments	Trimethyl ammonio, triethyl ammonio, diethyl amino and Sulfo derivatives of guar gum	Adsorbent
Waste water purifier	Cationic guar gum derivatives	Flocculating and exchanging agent
15. Analytical		
Hydrophylic polysaccharide matrix	Guar gum derivatives (Glycine hydroxamate in guaran, acetic acid hydroxamate in guaran, imino diacetic acid dihydroxamate in guaran)	Separation of metal ions

Support for affinity chromatography	Modified guar gum	Support for immobiliation of ligands
Liquid crystal	Guar gum	Thickener
Bioassay of calcium	Guar gum	Increases acceptability to rat
Beads	Guar gum	Purificationof lectins
Gel filtration	Cross linked guar gum	Chromatographic separation
Selective resin	Cross linked guar gum	Resin for separation of boron
16. Photography		
Processing solution	Hydrolysed guar gum	Binder
17. Shoe industry		
Guar gum	Guar gum	Improves the properties of the substrate through better fibrllation & reaggeregation of the collagen fibres
18. Food		
Modified starch with gum and salt	Guar gum	Thickener
Modifier of wheat flour	Guar gum	Modify visco-elastic behaviour of wheat flour
Sauce	Guar gum	Stabilizer
Myannaise & sauces	Guar gum	Thickening agent
Dia-mel diet preserves	Guar gum	Gelling agent
Ice-cream	Guar gum	Thickener
Ice-cream	Guar gum with carragenan and O-carboxy methyl cellulose	Stable thixotropic stabilizer, emulsifier system
Ice-cream	Guar gum with carboxymethyl cellulose and tamarind seed gum	Thickening and improver
Ice-cream	Guar gum with carrageenan gum	Stabilization
Potato agglomerate	Guar gum	Mucus forming property
Sugar purification	Glycidyl trimethyl ammonium denatured guar gum	Improves sedimentation
Canned products	Xanthan gum, guar gum, Carob gum	Thickening agent
Cheese	Guar gum with xanthan and carob gum	Stabilizer
Icing stabilizers	Guar gum	Stabilizer

Protein low bread	Guar gum	Binder
Bakery filling	Guar gum	Freeze thaw and heat stabilizer
Fat containing whips	Guar gum with whipable albumin and edible fat	To create pudding & cream dressing of cakes
Liquid egg	Guar gum with or without xanthan gum	Improves consistency
19. Dairy		
Fortified milk	Guar gum	Calcium binder
Whey protein concentrate	Guar gum	Thickener
20. Pharmaceutical/medicinal		
Monolithic polymeric matrix	Sodium carboxymethyl guar	Binds medicine
Matrix formulation of tablet	Guar gum	Binder
Compression coat	Guar gum	Drug targeting to colon
Tablets	modified guar gum	Sustained release of diltiazem-hydrochloride
Tablet formation	Guar gum/starch combination	—
Tablet additive	Guar gum	Disintegrant or binder
Tablet prep	Guar gum	Dry binder
Guar gum	Guar gum	Improves blood lipid blood pressure and fibrinolysis
Anticholest	Guar gum	Cholesterol lowering agent
PHGG	Partially hydrolysed guar gum with β-D-mannanase	Serum cholesterol decreasing agent
Dietary fibre	Partially hydrolysed guar gum	Blood glucose &blood lipid lowering agent
Soluble fibre	Guar gum	Improves glucose tolerance
Lactulose guar gum	Guar gum	Decreases transit time of colon
Guar pasta	Guar gum	Causes sharp decrease in glycaemic value
Molded mixture of pullulan and guar gum	Guar gum	Slow release of drug

Medicine for lowering glucose level of glyburide	Guar gum	Insulinogenic & blood glucose lowering agent
Anti-inflammatory drug	Guar gum	Suspending agent
Dietatic comp	Guar gum	Malnutrition treatment
For gastro intestinal disorders	Guar gum	—
Micro encapsulation of drugs	Guar gum	High resorptivity
Preparation for treatment of gastrointestinal ulcers and diarrhoea	Guar gum	Synergistic activity with bismuth salt
Vit. B12 prep.	Guar gum	Stabilizing
21. Agriculture		
Soil conditioner	Guar gum or Guar gum + polyvinyl alcohol + Borax or guar gum + borax	Increased water retention capacity of soil
Super absorbent polymer (soil additive)	Guar gum derivative	Improves water holding capacity
Adhesive for inoculating azotobacter	Guar gum	Adhesive of azotobacter
Anti crusting agent	Guar gum	Anticrusting agent
Water percolation preventor	Guar gum polyacrylamide	Water retaining agent
Poultry diet	Guar gum	Decreasing cholesterol level, decreases nitrogen retention, fat absorption and metabolizable energy
Coating for fertilizer	Guar gum with fatty acid derivatives and kaolin	Prevention of granules
Insecticidal composition	Guar gum	—
22. Cosmetics		
Anti dandruff cream	Guar gum with hydroxyethylcellulose and Zn pyrithione etc.	Suspension agent
Hair setting	Guar gum/urea and sulphite	Thickener
23. Soap industry		
Gel shampoo	Hydroxy-G-6 alkyl ether derivatives of guar gum	Thickener
Cleansing composition	Guar gum	Removes soil but retains lipid
Liquid detergent	Hydroxy propyl guar gum	Thickener

Skin cleaner	Guar gum	Thickener
24. Miscellaneous		
Water swellable clay composition	Guar gum	Water absorber and binding agent
Kojak gum substitute	Guar gum	Stabilizer
Tooth paste	Carboxy methylated guar gum	Thickening agent
Ink		Thickener
Heat storage composition	Hydroxy propyl ether of guar gum	Super coolable heat storage
Retracting, insulating, annuing composition	Jaguar HP guar gum	Water absorber
Chalk for marking on stone	Guar gum	Binder
Civil disorder control	Guar gum	Lubricant for concrete surface
Turbulent flow between pipe walls	Guar gum	Reduction of resistance
Aluminium reflectors	Guar gum	In increasing reflecting power
Production of sodium carbonate	Substituted guar gum	Clarifier

GUAR MEAL

The germ and outer seed coat of guar seed together constitute guar meal. Removal of gum from guar seeds increased the protein content of the residual byproduct i.e. guar meal. It is light, greyish tan material with beany flavour. The guar seeds results in 62-68% of meal having a protein content of 35-46%.

Chemical Composition

Proximate composition

Guar meal is a rich source of protein. It contains about one and half times more protein than guar seed and compares well with other vegetable protein source like oilseed cake used in poultry diets. The proximate composition of guar seed and its by-product has been reported by various workers (Table 5). From time to time various workers have tried to improve its nutritive value. Kaur *et al.* (1981) reported proximate composition and nutritive value of defatted guar meal, protein isolate and protein concentrate (Table 5) and found that protein concentrate is best for monogastric animals. Nagar *et al.* (1981 found that autoclaving of toasted guar meal did not change its crude protein content, however, acid treatment increased the crude protein content. Rajput *et al.* (1998) observed that the total protein of defatted guar meal consisted of 4 fractions, two of high and other two of low molecular weight proteins and detoxification of guar meal did not cause any significant change in the protein while autoclaving resulted in drastic changes. Joshi *et al.* (1990) compared the chemical composition of *Cyamopsis tetragonoloba* (L.) Taub. with *C. serrata* Schinze and *C. senegelensis* and found that *C. tetragonoloba* containd maximum protein and oil contents (Table 6).

Table 5. Chemical composition of guar meal and guar seed (per cent dry matter)

Product Name	Crude fibre	Crude protein	Nitr. free extract	Ether extract	Ash	Gum content	Haem olytic test	Reference
Guar seed	7.77	24.62	59.58	3.84	4.19	α	α	Sen and Ray, 1964
	10.52	29.43	–	5.24	3.84	α	α	Nagpal *et al.* 1971
Guar meal	8.42	46.92	34.13	5.20	5.33	α	α	Srivastava& Singh, 1960
	6.70	39.50	43.30	5.50	5.00	α	α	Patel *et al* 1970
	8.27	43.00	23.07	5.78	5.18	α	α	Narang, 1971
Defatted guar meal	9.20	51.62	17.26	4.24	4.30	13.27	++++	Kaur *et al.*, 1981
Guar protein isolate	1.40	72.61	14.59	–	2.90	8.50	- - -+	Kaur *et al.*, 1981
Guar protein concentrate	10.90	64.31	8.90	–	3.40	12.41	- - -	Kaur *et al.*, 1981

Source : Arora and Joshi (1991), α = Not analysed.

Table 6. Chemical composition of seeds of three species of *Cyamopsis*

Species	Gum (%)	Protein (%)	Oil (%)	100 seed weight (g)
C. tetragonoloba	31.30	32.58	6.1-7.5	2.99
C. serrata	30.25	27.12	4.40	0.90
C. senegelensis	30.60	26.25	3.06	0.54

Joshi *et al.* (1990).

Amino acid composition

Several workers have reported the amino acid composition of guar and its byproducts which speak about the quality of protein. Rama Krishanan (1957) reported that guar seed and guar meal were good source of essential amino acids (Table 7). Nagpal (1968) reported that guar meal have sufficient amount of lysine and histidine, tryptophan content was fair but methionine was deficient. Ambegaokar *et al.* (1969) analysed guar meal for essential amino acids and observed that tryptophan was the first limiting amino acid followed by methionine and threonine. Amino acid profile in general showed that guar seed was low in methionine content. However, the reported values for methionine as well as for other amino acids differed widely (Table 7).

Table 7. Available amino acids of guar meal and its products

Amino acids	Free amino acid (g/16 gN)		Guar meal		Dufatted guar meal (g/16 gN)	Guar protein (g/16 g N)	Guar protein concentrate (g/16gN)
	Guar (crude)	48 hrs germinated	Coarse (% food)	Fine (%) food			
Cystein	–	0.80	–		1.00	1.21	0.88
Lysine	9.10	12.30	3.05	2.43	3.95	4.14	4.30
Histidine	0.10	0.14	2.54	1.30	2.62	2.90	2.75
Arginine	1.20	2.40	7.61	4.80	13.49	14.70	15.19
Aspartic acid	–	12.20			9.93	12.05	9.92
Serine	–	1.80			4.42	4.70	4.52
Glycine	0.07	1.23			7.66	7.15	7.36
Glutamic acid	15.20	24.20			20.10	18.59	20.26
Threonine	–	1.90			2.67	3.84	2.83
Alanine	0.12	1.20			3.08	3.65	3.42
Tryptophane	0.10	0.70	0.49	0.36	0.88	1.13	0.82
Tyrosine	0.50	1.20			3.73	3.94	2.77
Valine	5.30	7.50			3.07	2.83	3.29
Methionine	3.10	5.60	0.34	0.34	1.16	1.07	1.27
Phenylalanine	–	0.30			3.80	4.32	3.73
Leucine	8.20	13.20	3.06		5.84	6.11	5.75
Isoleucine	2.10	1080			2.90	3.12	3.26
	Rama Krishanan (1957)		Nagpal (1986)		Kaur *et al.* (1981)	Kaur *et al.* (1981)	Kaur *et al.* (1981)

Source : Arora and Joshi (1991).

Oil and its fatty acid composition

Indian production of guar seed is more than 1200 thousand tonnes per year and it contains 5-6 per cent oil in the seed. Thus, total oil from guar seed could be extracted to an extent of approximately 60 thousands tonnes (Arora *et al.*, 1991). Nagpal *et al.* (1971) and Narang (1971) have reported that the guar meal contained 3.4 to 5.7 per cent ether extract. Joshi *et. al.* (1981) reported the fatty acid composition and physical constants of guar oil, whereas Arora *et al.* (1985) observed that the ratio of polar and non-polar lipids varied with the cultivars, it was 1:3 in HG 75 and 1:2 in HG 182 (Table 8).

Table 8. Fatty acid composition (relative %) of oil and its components (polar and non-polar lipids) in promising varieties of guar

Fatty acids	HG 75 lipids			HG 182 lipids		
	Total	Polar	Non-polar	Total	Polar	Non-polar
Palmitic	24.97	28.03	24.97	28.02	28.13	27.12
Stearic	5.55	6.81	6.66	7.01	6.25	5.71
Oleic	23.59	27.27	21.64	19.85	23.44	18.55
Linoleic	38.85	34.08	41.62	39.22	35.93	41.39
Linolenic	6.84	3.79	4.99	5.84	6.25	7.14
Total saturated	30.52	34.84	31.63	35.03	34.38	32.83
Total unsaturated fatty acids	69.38	65.14	68.25	64.91	65.62	67.08

Arora *et al.*, (1985)

Joshi *et al.* (1981) observed that the guar oil contained 55.1 per cent linoleic acid as against 51.8 per cent in sunflower oil. The total unsaturated fatty acids were 78.7 and 92.0 per cent in guar and sunflower oil, respectively. Guar oil contained 3.36 per cent linolenic acid. The iodine value and refractive index of the guar oil is well comparable with that of sunflower oil. Singh and Misra (1981) also reported that fatty acid composition of guar seed oil was generally similar to that of common edible oils. Three major species of Cyamopsis viz., *tetragonoloba, senegalensis* and *serrata* contained 6.1, 4.4 and 3.6 per cent oil, respectively (Arora *et al.*, 1991). The unsaturated fatty acids were highest in *serrata* and fatty acid composition was also comparable with that of sunflower oil.

Toxic constituents

Toxic constituents are commonly present in leguminous plants. Many investigations have been directed towards the determination of nature of toxic constituents present in guar and guar meal from time to time, which have been briefed below:

(i) Guar gum

There are reports that the residual gum has deleterious effects in chicks. The major cause of inhibition in the utilisation of guar meal is the residual gum. As low as 1 per cent guar gum in chicken diets caused a definite growth depression. As the level of guar gum was decreased to 0.5 or 0.25 per cent, the growth inhibition was also proportionately reduced. Bakshi *et al*. (1964) reported that toxicity of guar meal may be due to residual gum which is present to an extent of 18 per cent in guar meal. This gum is polysaccharide and is apparently not digested and absorbed from gastro-intestinal tract and thus, causes stickyness of droppings. When chicks were given 1.8 per cent or 2.7 per cent guar gum, growth was depressed. The gum seemed not to be digested and proportion of gum in guar meal may restrict its amount used in diets.

(ii) Trypsin inhibitor and other toxic factors

Bakshi *et al*. (1964) reported that the ill effect of guar meal in chicken and hens was due to heat labile tryptic inhibitor which is macromolecule possibly a protein. Ahuja and Chopra (1975) reported that raw guar dhâl or guar dhal treated with saline water and supplemented with trypsin at 1 per cent in diet did not prevent growth depression in rats caused by guar dhal. This indicated that guar may contain other antigrowth factors beside, trypsin inhibitor. It is evident from the literature that toxic constituents are present in guar meal but the exact nature of the toxic constituents have not been reported. Nagra *et al*. (1994) reported that trypsin inhibitor activity in toasted guar meal (278 units) was decreased to 94 units/g due to autoclaving of the meal and was further lowered to 23-24 units/g through fermentation with non-toxic fungi i.e. *Aspergillus niger* and *Fusarium*.

Vidyasagar *et al*. (1978) reported the presence of toxic factors in guar meal and they tried to detoxify them. Gupta *et al.* (1978) reported that guar meal contains two deleterious factors i.e. residual guar gum and trypsin inhibitor and toasting of guar meal improves its nutritive value. However, Arora and Joshi (1980) were able to conclude the other toxic factor saponian in guar meal, which in thermostable. The extracted guar saponin foamed easily when shaken in aqueous solution and it was strongly haemolytic and varied from 10-12 per cent in guar meal. The nature of saponin in guar is of different type than that of alfalfa and soybean (Fig. 3).

The toxic nature of the crude extract prepared from raw guar meal (defatted) have been confirmed by injecting it, intraperitonially to the rats and mice. Nutritional studies indicates the presence of another growth depressing factor beside saponin. The purified saponin extract is not so much toxic as crude extract. The LD_{50} value for rats were 750, 460 and 300 mg/kg body weight for 24, 48 and 72 hours whereas, for mice the same were 1500, 780, 780 mg/kg body weight for 24, 48 and 72 hours, respectively, (Joshi *et al.*, 1985). Rajput *et al.* (1998) reported that autoclaving of raw guar meal resulted in considerable destruction of haemagglutinins (84%) and trypsin inhibitors (84%) without much changing the saponin (6.1%) and phytate (7.5%) content. Further, acidified water leaching of autoclaved meal decreased the concentration of saponin to 2.1 per cent, phytate to 3.9 per cent and polyphenols to 0.76 per cent. Extracting the meal with aqueous alcohols lowered the concentration of saponin and haemagglutinins to nil. Earlier,

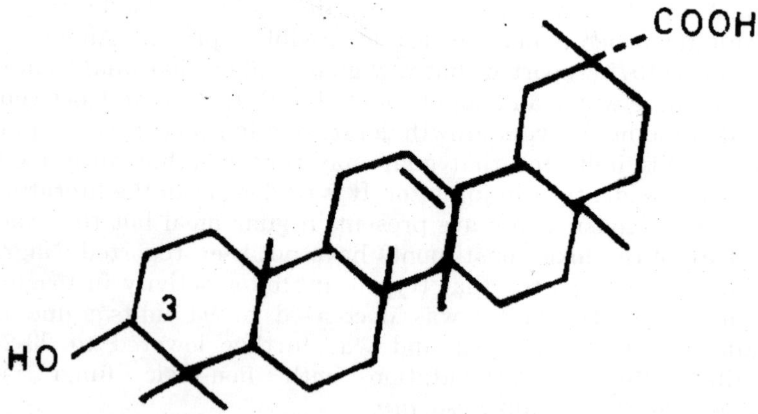

Fig. 3 Epikatonic acid

Nagpal *et al.* (1971) reported phenolic compounds to be responsible for the toxicity of guar meal, Kaushal and Bhatia (1982) identified gallotanins, gallic acid, gallicacid derivatives, myricetin-7 glucoside-3-glycoside, Kaempferol-7 glucoside 3-glucoside, Kaempferol-3-rutinoside, Kaempferol-3 glucoside, chlorogenic acid, caffeic acid and ellagic acid from guar seed.

Possible role of seed coat polyphenolics on the trypsin inhibitor activity of guar was also studied. The trypsin inhibitor activity in guar was found to be heat stable, and significantly correlated with tannic acid concentration of the seed coat, though seed coat devoid of soluble protein is having maximum trypsin inhibitor activity. It was concluded that tannin plays an important role as trypsin inhibitor than the heat labile proteinaceous trypsin inhibitor in guar (Jain *et al.*, 1983).

NUTRITIVE VALUE OF GUAR MEAL

To assess the nutritive value of guar meal, feeding trials have been conducted by various workers in rodents, poultry and ruminants. Attempts have also been made to improve its nutritive value and studied the impact of guar meal on the transport of amino acid.

Rodents feeding trials

Feeding experiments conducted by various workers indicate poor growth with guar meal. However, different treatments resulted in improvement of guar meal, as livestock feed. Kawatra *et al.* (1969) made different treatments on guar meal, (i) acid treatments, (ii) hot water treatment, (iii) enzyme treatment to detoxify guar-meal. Extraction with boiling water and treatment with hydrochloric acid resulted in an increase in the PER values of guar meal at 10 per cent protein level from 0.16 to 0.96 and 1.25, respectively. This improvement in the nutritive value of the proteins was reflected in increase in nitrogen content in liver and blood plasma. The postmortem examination of rats showed that the liver and other organs looked normal, except that the intestine was filled with gases and dark fluid which appeared haemorrhagic. Plasma protein of rats fed guar dhal (meal) with or without added trypsin were significantly less than in casein group. Ahuja and Chopra (1980) separated two major fractions of protein. Fraction B was found to have a very high trypsin inhibiting activity, whereas, fraction A had little. Mice fed fraction A as protein source at 10 and 15 per cent levels were able to maintain growth but those fed on fraction B at 5 per cent level were unable to survive. *In vitro* digestibility of fractions B, A and casein by trypsin and pancreatin at 37°C increased progressively in that order. Kaur *et al.* (1981) conducted feeding trials of rats with defatted guar meal, protein isolate and protein concentrate and observed 60-80 per cent mortality of rats with defatted guar meal. Diets containing defatted guar meal and guar protein isolate, having biological value of 47.17 and 53.72, gave negative PER value. Protein concentrate having the biological value as of defatted meal, gave PER of 1.98 as against 2.5 with casein. Amongst the three preparations, protein concentrate was good nutritionally and is free from toxic factors.

Poultry feeding trials

The lower concentration of guar meal in the diet has been found satisfactory,while higher percentage decreased the feed efficiency. Sathe and Bose (1962) reported that use of guar meal at 12.5 per cent level in all mash growing rations had an adverse effect on feed efficiency. At 6.25 per cent, growth rate was maintained but feed efficiency was lowered. The guar meal caused loose sticky type diarrhoea and growth depression in chicks. Bakshi (1967) found that when guar meal was fed to chickens it increased the size of pancreas but had no adverse effect on heart, liver or kidney. Nagpal (1968) found that guar meal and autoclaved guar meal when used as a sole source of protein in otherwise adequate diet resulted in emaciation of chicks and heavy mortality during feeding period of 24 days. Thakur and Pradhan (1975) studied inclusion of guar meal at different levels and found that it had a deleterious effect on dressed weight, eviscerated weight and ready to cook meat. Chemical analysis of breast and thigh meat showed that protein and ether extract content increased due to increased dietary protein level and decreased metabolizable energy, due to guar meal inclusion in diet. Pradhan *et al.* (1978) conducted feeding trials with different levels of toasted guar meal (GMT), raw guar meal (GMR) and control diet with broiler chicks and observed that the chicks receiving guar meal in their diets irrespective of raw or toasted levels, consumed significantly ($P<0.05$) more feed than control. Feed efficiency was adversely affected by all the diets with guar meal except diets with 5 to 10 per cent GMT. Feed efficiency with 5 per cent GMR and 5 to 10 per cent GMT diets was as good as control. Mortality rate with 15 per cent GMR diet was as high as 62 per cent. They concluded that toasting process improved the utilization of guar meal upto 10 per cent and also reduced the chick mortality.

Ruminants feeding trials

Branford and Sewell (1922) were the first workers in India to examine the feeding value of guar in livestock farm, Hisar. There was a local superstition that guar is not a good feed in hot weather. It was said to be too heating. They compared the nutritive value of gram, til cake, mustard cake, mixed concentrate and guar in young and working bullocks. The mixed concentrate contained equal parts of oat, gram, guar and bran. Til cake was found to be the best concentrate for growth followed in descending order of nutritive value by mustard cake, gram, guar and mixed concentrates. Dasgupta (1945) conducted series of trials in growing heifers to study the relative feeding values of five oil cakes viz., linseed cack, decorticated and undecorticated groundnut cake, rapeseed cake and til cake of 4 legume seeds and their by products viz. guar meal, gram, arhar and moong churi. Moong churi gave best results. Sadagopan and Tàlaptra (1961) reported that if guar meal is used as a sole concentrate, it was liable to cause diarrhoea in animals.

Improvement of feeding value

Several attempts have been made by various investigators to detoxify guar meal with and without success and to make it suitable for monogastric animals feed. Kawatra (1965) reported that cooking increased the palatability of the guar meal which was evidenced by high consumption of toasted guar meal mash as compared to that of untreated guar meal mash. However, cooking or disodium phosphate was not able to detoxify guar meal. Manuja (1966) reported that addition of germinating legumes to chick rations containing guar meal did not improve its nutritive value. Bakshi (1967) reported that the growth depressing effect of guar meal can be prevented by suitable treatment of meal at 110°C. Enzyme pectinase when incorporated in diet increased weight gain. Ambegaokar *et al.* (1969) could not find any heat labile growth inhibitor in raw guar meal. Fortification of autoclaved guar meal with 1 per cent DL-methionine did not result in any significant improvement in the growth promoting rate of the diet. Kawatra *et al.* (1969) reported that enzyme treatment did not result in any improvement in the nutritive value of guar meal for rats. Extraction with boiling water and HCl treatment resulted in increase in the protein efficiency ratio of the raw guar meal which was further indicated by increased nitrogen content of liver and blood plasma. Kaur *et al*. (1981) also detoxified the guar meal and found that protein concentrate was good nutritionally and was free from toxic factors.

In addition to various industrial uses of gum, guar holds great promise for supplying much needed protein in the human diet. Guar protein as germ is presently mixed with seed coat, sole as cattle feed and after detoxifying are recommended for poultry ration. Guar protein could add significantly to the world's protein supply, providing increased vegetable protein for human dietary needs.

Man's earliest use of guar as food was to cook fresh green pods after cutting and often with potatoes. The immature pods are sometimes dried, salted and preserved for future use. The dried pods may also be fried like potato chips. More mature green pod are often cooked like French beans (Chopra *et al*., 1956).

Utilization of guar meal

Kaur *et al*. (1981) found that the protein concentrate prepared is free from toxic constituents and beany flavour, Beany flavour or bitterness in guar meal has been a limiting factor for its unpopularity for human consumption. However, the protein concentrate prepared from guar meal is free from beany flavour and author used protein concentrate to fortify the biscuits which were more tasty and crispy than "vitamino" biscuits developed by CFTRI, Mysore. This has opened up new avenues for utilization of guar meal for human consumption and will have an important bearing on the economics and viability of guar industry in the long run.

FUTURE THRUST

Guar a multipurpose crop still have wide scope and future thrust, the important ones are listed below :

(i) There is still scope to explore more uses of guar gum. Since guar gum consists of galactose, so it could be used to control diabeties and more research is required in this direction. Further, it could be a boon for diabetic and heart patient too (since guar gum also helps in lowering the cholesterol level in the blood). The industry and research staff should cooperate in this regard and instead of exporting raw guar gum more products should be prepared and these could be exported to earn more foreign exchange.

(ii) Guar meal a byproduct of guar gum industry could be used in the diet of ruminants and can replace other protein supplements like ground nut oil cake. Further, toasted guar meal could be used upto 5-10 per cent in poultry rations. The use of protein concentrate (prepared from guar meal after removing toxic factor) for fortification of biscuits has opened up new avenues for utilization of guar meal for human consumption. Thus, in future procedure could be standardised for economic detoxification of guar meal which could be used for feeding ruminants, poultry and humans as a protein source.

SUMMARY

Guar is a multipurpose crop, the seed consists of galactomannan polysaccharide gum in endosperm, called as guar gum. Guar gum is an agroindustrial product having innumerable end uses and helps the country to earn the foreign exchange. Its chemistry, properties and industrial applications have been discussed. The remaining part of seed a byproduct of guar gum industry is called as guar meal. It is light grayish tan material with beany flavour. It is rich source of protein and essential amino acids but limiting in methionine, tryptophane and threonine. It consist of 5-6 per cent oil and fatty acid composition of guar oil is well comparable to that of common edible oils. Residual guar gum, trypsin inhibitor, polyphenols and saponins are the major toxic factors, present in guar meal. These toxic factors could be removed by processing and toxin free guar protein concentrate could be used for human and animal feeding. Guar pods and empty pods could also be used for human and animal feeding, respectively.

REFERENCES

Ahuja, B.S. and Chopra, A.K. 1975. Studies on the biological evaluation of guar (*Cyamopsis tetragonoloba*). Effect of trypsin supplementation on the nutritive value of guar. *Indian J. Nutr. Diet.* **12**: 211-15.

Ahuja, B.S. and Chopra, A.K. 1980. Evaluation of guar protein fractions for their nutritive value. *Baroda J. Nutr.* **7**: 7.

Ambegaokar, S.D., Kameth, J.K. and Shinde, W.P. 1969. Nutritional studies in protein guar (*Cyamopsis tetragonoloba*). *Indian J. Nutr. Diet.* **6**: 323.

APEDA 1999. A study on guar gum. Management Consultancy Division Dalal Consultant and Engineers Ltd. 501, Sakar II, Ellisbridge, Ahmedabad. pp. 22-35.

Arora, S.K. 1981. Guar - its uses are many and expanding all the times. Presented at Fourth All India Guar Res. Dev. Workshop held at CAZRI, Jodhpur on March, 10-11.

Arora, S.K. 1989. Guar Endosperm – Its Chemistry and Utilization. In : Indian Agro Exports, S.S. Jasol (ed.), pp 277-302.

Arora, S.K. and Joshi, U.N. 1980. A thermostable haemolytic factor in guar (*Cyamopsis tetragonoloba*). *Guar Newsl.* 1 : 13-14.

Arora, S.K., Joshi, U.N. and Arora, R.N. 1991. Fatty acid spectrum of guar species. *Guar Res. Annals.* **7**: 57-59.

Arora, S.K., Joshi, U.N. and Jain, V. 1985. Lipid classes and fatty acid composition of two promising varieties of *Cyamopsis tetragonoloba* (L.) Taub. *Guar Res. Annals* **4**: 9-10.

Arora, S.K., Paroda, R.S. and Saini, M.L. 1980. Chemical composition and *in vitro* dry matter digestibility of empty guar pods. *Guar Newsletter.* 1 : 10-11.

Bakshi, Y.K. 1967. Studies on toxicity and processing of guar meal. Dissertation Abstr. (B) **27**: 2268.

Bakshi, Y.K., Greger, C.R. and Couch, J.R. 1964. Studies on guar meal. *Poult. Sci.* **43**: 1302.

Branford, R. and Sewell, E. 1922. Comparative feeding value of common protein supplements. *Bull. Agric. Res. Inst. Pusa,* **130**: 1.

Chopra, R.N., Nayar, S.L. and Chopra, I.C. 1956. Glossary of Indian Medicinal Plants, Ranchi, Catholic Press.

Das, B. and Arora, S.K. 1978. Guar seed – Its chemistry and industrial utilization of gum. In: Guar-Its Improvement and Management (eds. R.S. Paroda and S.K. Arora), *Forage Res.* **4**(A) pp. 79-101. The Indian Society of Forage Research, Hisar, India.

Das, B., Arora, S.K. and Luthra, Y.P. 1977. A rapid method for determination of gum in guar (*Cyamopsis tetragonoloba* (L.) Taub.). Proc. First ICAR Guar Res. Workshop, Central Arid Zone Res. Inst., Jodhpur (India). P. 117-23 Jan. 11 and 12th.

Dasgupta, N.C. 1945. The comparative value of some concentrates in the feed of growing cattle. *Indian J. Vet. Sci. Anim. Husb.* **15**: 123-133.

Deb, S.K. and Mukerjee, S.N. 1963. Molecular weight and dimensions of guar gum from light scattering in solution. *Indian J. Chem.* **1**: 413-14.

Gupta, S.L. and Bhatia, R.K.S. 1970. Hysteresis in water vapour sorption and desorption on guar gum. *J. Indian Chem. Soc.* **47**: 1203-204.

Gupta, P.C., Vidya Sagar and Pradhan, K. 1978. Guar as feed and fodder crop. In: Guar-its improvement and management (eds. Paroda, R.S. and Arora, S.K.) *Forage Res.* **4A**: 109-22.

Jain, V., Joshi, U.N. and Arora, S.K. 1983. Possible role of seed coat polyphenolics on the trypsin inhibitor activity of guar. Presented at Third Indian Convention of Food Science and Technology, from June, 2-4 held at Mysore.

Joshi, U.N. and Arora, S.K. 1993. Recent advances in guar gum chemistry and utilization – A review. *Forage Res.* **19**(3&4): 310-28.

Joshi, U.N., Arora, S.K. and Arora, R.N. 1990. Differential chemical composition of guar species. *Guar Res. Annals* **6**: 38-40.

Joshi, U.N., Arora, S.K. and Singh, J.V. 1991. Carbohydrate make up and nutritive value of guar (*Cyamopsis tetragonoloba* (L.) Taub.). *Guar Res. Ann.* **7**: 48-53.

Joshi, U.N., Arora, S.K., Verma, S.P. and Jain, V. 1985. LD_{50} of crude growth depressing factor of *Cyamopsis tetragonoloba* (L.) Taub. *Guar Res. Ann.* **4**: 11-12.

Joshi, U.N., Kaur, A. and Arora, S.K. 1981. Fatty acid composition of guar oil. *Guar News Letter.* **2**: 43-44.

Kaushal, G.P. and Bhatia, I.S. 1982. A study of polyphenols in the seeds and leaves of guar *Cyamopsis tetragonoloba* (L.) Taub. *J. Sci. Food Agric.* **33**(5): 461-70.

Kaur, A., Arora, S.K. and Joshi, U.N. 1981. Nutritive value of guar meal, protein isolate and concentrate. *Guar News Letter* **2**: 41-43.

Kaur, A., Arora, S.K. and Joshi, U.N. 1981. Amino acid composition of protein isolate and concentrate of guar meal. *Guar News Letter* **2**: 44-7.

Kawatra, B.L. 1965. Research on processing of guar meal and castor oil cake for use as livestock feed. M.Sc. Thesis, Punjab Agril. Univ., Ludhiana.

Kawatra, B.L., Garcha, J.S. and Wagle, D.S. 1969. Studies on guar (*Cyamopsis tetragonoloba*) protein. *Indian J. Nutr. Dietet.* **6**: 91.

Kawatra, B.L., Garcha, J.S. and Wagle, D.S. 1971. Effect of feeding guar gum and of guar meal on N-balance. *J. Res. P.A.U.* 8 : 227-231.

Kendurkar, S.G. 1970. Refining guar gum. *Indian Patent.* **114**: 180 (Cf. *Chem. Abstr.* **74**: 1279514, 1971).

Manuja, N.K. 1966. Studies on the detoxification of guar meal.

M.Sc. Thesis, Punjab Agril. Univ., Hisar.

Mehta, D.R. and Ramakrishanan, C.V. 1957. Studies on guar seed oil (*Cyamopsis psoralioides*). *J. Am. Soc. Oil Chem.* **34**: 459-61.

Mehrotra, D.N., Tripathi, R.D., Srivastava, G.P. and Misra, M.C. 1975. Food value of some guar (*Cyamopsis psoralioides*) varieties. *Indian J. Agric. Res.* **9**: 37-42.

Mital, S.P., Thomas, T.A., Dabas, B.S. and Lal, B.M. 1971. Gum content as related to seed yield and other characters in guar. *Indian J. Gen. Pl. Breed.* **31**: 228-32.

Nagar, S.S., Singavi, B.K. and Ichponani, J.S. 1981. Chemical composition of fractions processed from guar meal. *Indian J. Poult. Sci.* **16**: 259-62.

Nagpal, M.L. 1968. Studies on the chemical and biological evaluation of some poultry feed stuffs with special reference to guar meal (*Cyamopsis tetragonoloba*). Ph.D. Dissertation, Punjab Agril. Univ., Ludhiana.

Nagpal, M.L., Agarwal, O.P. and Bhatia, I.S. 1971. Chemical and biological examination of guar meal (*Cyamopsis tetragonoloba). Indian J. Anim. Sci.* **41**: 283-93.

Nagra, S.S., Sethi, R.P., Chawla, J.S. and Chopra, A.K. 1994. Improvement in nutritional value of guar meal by fungal fermentation. *Indian J. Animal Nutr.* **11**(1): 7-11.

Narang, M.P. 1971. Effect of different sources of dietary nitrogen on rumen liquor and blood of buffaloes. Ph.D. Dissertation, H.A.U., Hisar.

Patel, R.R. 1958. *Gum Indian Patent.* **61**: 005. (*Cf. Chem. Abstr.* **52**: 21190c, 1958).

Patel, R.R. 1958b. *Gum. Indian Patent* 61 : 044. (*Cf. Chem. Abstr.* 52 : 21990c, 1958).

Pradhan, K., Vidya Sagar and Prasad, D. 1978. Effect of processing on the nutritional value of guar meal for poultry. A research report, Deptt. of Animal Nutrition, H.A.U., Hisar.

Rajput, L.P., Ramamani, S., Saleem, M.A. and Subramanian, N. 1998. Chemical and biological studies on processed guar (*Cyamopsis tetragonoloba*) meal. *Indian J. Poult. Sci.* **33**(1): 15-25.

Ramakrishanan, C.V. 1957. Amino acid composition of crude and germinated guar seed flour protein (*Cyamopsis psoralioides). Experimentia* **13**: 78.

Relan, P.S., Kumar, S., Garg, V. and Arora, S.K. 1991. Rheological studies of guar gum by viscometric method. *Guar Res. Ann.* **7**: 25-29.

Sadagopan, V.R. and Talapatra, S.K. 1961. Biological assay of guar meal on growth rate of Haryana calves. *Indian Vet. J.* **44**: 1061-68.

Sathe, B.S. and Bose, S. 1962. Studies on the utilization of industrial and farm byproducts in growing poultry rations. *Indian J. Vet. Sci.* **32**: 74-81.

Sehgal, K., Nainawatee, H.S. and Lal, B.M. 1973. Galactomannan degrading enzyme from germinating *Cyamopsis tetragonoloba* seeds. *Biochem. Physiol. Pflanz.* **164**: 423-28.

Srivastava, S.P. and Singh, G.S. 1960. Digestbility coefficient and nutritive value of guar meal or churi. *Indian J. Dairy Sci.* **13** : 136-142.

Shinde, S.K. and Bhargava, P.P. 1968. Fatty oil from the guar seed cotyledons. *Indian Oil Soap J.* **33**: 195-97.

Sikka, K.C. and Johari, R.P. 1969. Mannose containing polysaccharides. A method for the extraction and estimation of gum of guar (*Cyamopsis tetragonoloba*). *Res. Ind.* **14**: 138-39.

Singh, S.P. and Mishra, B.K. 1981. Lipids of guar seed meal (*Cyamopsis tetragonoloba*). *J. Agric. Food Chem.* **20**(5): 907-9.

Taghavi, S.M. and Vora, A.B. 1994. Effect of industrial effluent on germination and growth development of guar seed (var. PNB). *J. Environ. Biol.* **15(3)**: 209-212.

Sundarajan, P.R. and Rao, V.S.R. 1970. Conformational studies of linears ß-D-1,4 mannan and galactan. *Biopolymers.* **9**: 1239-47 (Cf. *Chem. Abstr.* **73**: 126972, 1970).

Thakur, R.S. and Pradhan, K. 1975. A note on inclusion of guar meal in broiler ration. *Indian J. Anim. Sci.* **45** : 98-102.

Vidya Sagar, Prasad, D., Thakur, R.S. and Pradhan, K. 1978. Nutritional evaluation of processed guar (*Cyamopsis tetragonoloba*) meal for broilers. *Indian J.Poult. Sci.* **13** (3) : 155-60.

CHAPTER - 11

INDUSTRIAL ASPECTS

N.S. Yadav and R.L. Ray

INTRODUCTION

For thousand years, guar has been grown in the Indian sub-continent for use, mainly as cattle fodder and to a small extent, as vegetable for human consumption. In less than 50 years, it has been widely adopted, as a water-soluble natural polymer, whose unique properties have been utilised to produce a variety of products for application in many industries. Guar is now-a-days an important industrial crop both in India and Pakistan. It is a significant foreign exchange earner; because guar gum, its derivatives and the high protein meal produced from the seeds, are sold world over to an expanding market.

The word "Guar" is derived from Sanskrit word "Gau-ahar" and was used as cattle feed in India before 1953. It was in 1950's General Mill and Stein Hall and Co., USA, made break through to make it feasible in industry by manufacturing and replacing other gums due to its cold water solubility and versatility in end uses.

In India, in 1955, Birla Brothers started firstever guar, gum production by the name of "Bhiwani Gum and Guar Factory" in collaboration with Stein Hall, USA, the former was renamed as "Hindustan Gum and Chemicals Ltd. in 1962. Now there are many guar gum industries in India, which are playing a great role in domestic as well as in export market. It is placeable here that India and Pakistan, make 95% supply of guar gum in the world.

CHEMISTRY OF GUAR GUM

Guar gum is one of the principal components of the seed of the guar plant, *Cyamopsis tetragonoloba* L. contained in the endosperm.The guar is composed of hull, germ and endosperm parts. Germ part is heavily protein rich and endosperm part is mainly galactomannan rich. A typical composition of guar seed is as follows:-

Guar gum, a D-galactose-D- mannoglycan (1:2), is straight chain mannose with single membered galactose branches. The D-mannopyranose units are joined by Beeta-(1-4) linkages, and single D-galactopyranose units are joined to this chain by alpha (1-6) linkages. On an average, the galactose branches occur on every alternate mannose units. The molecular weight has been reported to at least 220,000 daltons.

CULTIVATION OF GUAR

Though, India is tropical country, yet it has Thar desert covering India and Pakistan. Thus, from Haryana to south Rajasthan, north Gujarat and Kutch areas have identical climatic conditions, suitable for growing of guar crop. Guar suits to sandy soil and requires low rainfall with longer dry spell intervals to grow it properly with full bloom. After 10-20 days of first showers, guar saplings reach at the height of 3-5 inches. Within one and half month, when the beans of guar come out, it require another rainfall. During the period of 90-95 days, guar crop becomes mature and ripes.

In case of abnormal and untimely rains, guar seed deviates from it normal size, being smaller in size. On exposure of guar crop to heavy continuous rains, seeds turn black and dark blue and quality deteriorates. Hence, special measures are required to be followed to make the products from such seeds. Guar seed supplied by farmers are not completely pure and contain fine and over sized muds, husks, stone etc. It should be purified before making refined guar gum split, otherwise quality of the split will be deteriorated.

PROCESSING OF SEEDS

Guar seeds, collected from farmers or stock market, contain lot of foreign particles like, fine dust, mud, husk, stones, chaff, broken seeds etc. These foreign materials are separated out before processing.

The industries are now-a days using Stone Mill or, Pin Mill to crack the clean seeds and making the germs (rich in protein, it is used as cattle feed in large scale) free. Cracked splits are passed through heated Kiln keeping for almost 1.5 minutes at 100-115^0 C. Controlling of temperature in the Kiln is important for getting the good quality splits. After Kiln, heated cracked splits are passed through Dehusked Machine. Temperature of split should be dropped quickly to ambient temperature to resist heat shock on split which otherwise can damage the viscosity and reactivity of the splits.

GUAR GUM POWDER (Processing of Guar gum split)

Guar gum powder can be manufactured from refined guar gum splits by grinding through Attrition Mills, Hammer Mill or any suitable size reduction equipments followed by hydration of the splits. Hydration of splits, and particle size play an important role to controlling viscosity and hydration rate of guar gum powder solution. After dehusking, splits are hydrated allowing moisture content about 20-70%, thereafter hydrated splits are ground through conventional grinding equipments. Sieving is necessary for getting required particle size.

PROPERTIES

Solubility

Guar gum and its derivates are soluble in cold water/polar solvent and do not swell in non-polar solvent. Rate of hydration depends on the particle size of the product and the viscosity. Rate of hydration is slow for non-derivatized coarse guar gum powder and viscosity will develop slowly. However, fine powder hydrates quickly to develop viscosity. To make a quick solubility, stirring of the paste with mechanical stirrer is necessary. The rpm of the stirrer also controls the rate of hydration; higher the rpm of the stirrer, quicker is the hydration of the product.

Some derivatized guar gum, like hydroxy propyl guar gum, carboxymethyl guar gum hydrate quickly compared to non-derivatized guar gum. Degree of substitution (DS) of derivatized guar gum governs the rate of solubility. Lesser the D.S., lesser the solubility and solubility increases with increasing D.S. to certain limit, the solubility decreases once again. For better solubility, D.S. from 0.4 to 0.7 has been observed more effective.

There are some guar gum derivative, like hydroxy propyl guar gum or, its mix derivative which are soluble in the 50% mixture of nonpolar solvent and water and develop viscosity. In such situation solubility becomes the function of non-solvent content and degree of substitution. Addition of salts give the reduced hydration.

Viscosity Vs Concentration

Viscosity of guar gum solution increases with its concentration and behaves like non-newtonian fluid. If 1% guar gum solution gives 2000 cps (by Brroke-field Viscometer at 20 rpm at 25 $O^{O}C$ with spnle # 4), 20% of the same product will slow around 20,000 cps (approx. increase of 10 folds). In case of modified guar gum; increase in the viscosity with concentration is rather less compared to unmodified guar gum, depending on the molecular weight of the gum and degree of depolymerization. In some modified textile gum, it increases by about 1.5 folds with the increase in concentration of the gum.

Salt Compatibility

Compatibility of guar gum solution with salt depends on the ionic nature of the salt, its reactivity and also on the ionic nature of the guar gum; product. It may be mentioned that guar gum product is available with non-ionic, anionic and cationic charge. Divalent and trivalent salts are not compatible with anionic guar gum product, but they are compatible with non-ionic guar product.

Cross Linkage with Borax

Guar gum forms a gel with borax at alkaline pH, but this property is reversible. It develops viscosity with borax in acidic pH and gel in alkaline pH. The optimum pH range for gel formation is 7.5-10.5. It is believed that polysaccharides with numerous adjacent hydroxyl groups in this position can form three dimensional borated gels. In an alkaline medium, the borate ion can react at C_2 and C_3 of the mannose units and C_3 and C_4 of galactose side branches.

Stability of Guar Gum

Stability of guar gum depends on the good manufacturing practices (GMP). Without addition of any preservative, generally, it shows 48 hours stability in the solution. However, if the gum is microbiologically contaminated too much, degradation of viscosity starts very fast. Stability of the guar paste also depends on pH of the solution. In alkaline pH, paste is more stable than in acidic pH.

Addition of preservative in the product increases stability of the paste depending on the type and dose of the preservative used. For industrial application, there is a scope to add the preservative, but in case of food grade product, there is little scope for the addition of preservative. Generally 0.5 to 1.0% preservative gives stability of the paste more than a week. In the dry form, stability of the guar gum depends on the moisture content of the finish product as well as temperature and storage conditions. Moisture content more than 14% in the powder degrades quickly. Generally viscosity of the guar gum decreases slowly. High moisture content, exposure to air and high ambient temperature during storage accelerate the deterioration of the viscosity.

Synergistic effects of Guar Gum

Guar gum shows its synergistic properties with other gums like, *Xanthan* gum, locust bean gum, *Caragenan* gum. When guar gum is mixed with *Xanthan* gum in some definite proportions, it gives synergistic properties in viscosity, i.e., viscosity of blended *Xanthan* and guar is higher than the individual viscosity of the either. Industry people take advantages of such properties for making many food grade blended products.

APPLICATIONS OF GUAR GUM PRODUCTS

Food Uses

The use of gum guar can be envisaged in all systems where water is an important factor. Numerous foodstuff compositions require the use of colloids to modify the behaviours of the water in relation to other components. Guar gum accordingly acts as a binder, stabilizer or thickener in the following products.

In Frozen Food

Due to water holding properties and ability to inhibit ice crystal formation, guar gum is used as a binder and stabilizer in many frozen food preparations, ice cream, ice pop, water-ice, lollies, *sherbets* and iced deserts.

Guar gum together with sugar, fat and water-free milk components provides a suitable mixture for the preparation of ice cream, milk shakes and alcoholic milk drinks

In Baked Food

Guar gum is used as a moisture-retention agent. It increases shelf-life in industrial bread making. A guar gum solution added to the dough while kneading, increases, the yield considerably. The dough produced, presents a greater resilience and a drier, less flabby appearance. Gum improves the elasticity of the bread. Generally guar gum is used 0.1-0.2% of the total weight of the dry ingredients.

In Processed Cheese

Guar gum improves the texture in cold pack cheese products. It works as a moisture distribution control agent. When guar gum is incorporated, yield of the curd solid increases. Without use of whey, guar improves the texture of the final product. Guar blended with *Xanthan*/locust bean gum gives more synergistic effects to stabilize the processed cheese.

Guar gum is not approved by the European Economic Countries for processed cheese stabilization. In U.S.A. the FDA approves guar gum as an optional ingredient in cheese products. The use of guar gum is "allowed as a formulation aid, stabilizer and thickener at a level of 0.8%".

In Dairy Products

Guar gum when used alone or combined with other water soluble gums such as locust bean gum, *Xanthan* gum, gelatine or modified starches becomes an excellent stabilizer for dairy desserts and acidified dairy products such as yoghurt, kefir and quarg.

In Dressings and Sauces

Guar gum is unique cold-water dispersible, acid-resistant, free water binder in salad dressing, sauces, pickles and relishes. It functions as an emulsion stabilizer by increasing the viscosity of the aqueous phase reducing the separation between water and oil phases. A patent using guar exists for the manufacture of fat emulsion in the form of paste for use in various sauces, soups, gravies and ready meals in sauce. The product improves the flavour and consistency and imparts a creamy texture.

In Instant Mixes

Many hydrocolloids are used as stabilizer and gelling agents in packaged food mixes but guar offers more functional advantages than others. It brings body and mouth-feel to instant milk deserts, puddings and chiffon pies. Guar is a fast, cold dispersible, thickening and texturing agent used in instant sauces, a cold or hot thickening and suspending agent for powdered chocolate mixes.

In Meat Products, Canned Meats

Guar acts as a binder and lubricant in the manufacture of a variety of meat products such as saucages, stuffed meat products, sliced meat products and canned meat, such as, corned beef, luncheon meat, pet foods and baby foods. Free-water separation in canned meat products is prevented when guar is used as binder. The cold water solubility and hydration of guar gum permits easy processing while absorbing and binding excess water. The resulting meat is firmer and consistency in the can is better. Guar gum is used at the levels of 0.2 to 0.7% based on the water used in the formulation.

In Beverages

Guar gum is used as a viscosity control and suspending agent in fruit beverages and fruit nectars. It improves mouth-feel in sugarless low calorie drinks. Guar gum is useful to resist the breakdown during sterilization in low pH condition. As it is cold water soluble, it can be used in beverage processing plants. It acts as a good stabiliser together with locust bean gum in fruit toppings, and fruit or chocolate syrups. Guar gum is also used to stabilise carbonate and bicarbonate used in the production of carbonated drinks and medical beverages to prevent premature decomposition.

As Textile Thickners

Depolymerised guar gum, which is made by chemical modification, is suitable to use as textile thickners for. sizing, printing and finishing of cotton, rayon, silk, polyester, polyamide, polyacetate, acetate, triacetate and wool. Nature and degree of modification of guar gum depends on the type of fabrics, and other textile auxillaries to be used during processing. Depolymerised guar gum having different concentration is available in the market. Due to its excellent flow properties and film forming properties, guar has replaced starch and sodium alginate in many industries.

In textile printing, suitable modified guar products are required which run through machine without any trouble and have compatibility with all chemical auxiliaries and colours used during printing process. Modified guar is also easily washable from printing cloth after fixation of colour.

Concentration of gum is decided by textile house based on the type of cloth to be used. For silk printing, they require very high concentration of the stock paste, whereas, for printing on polestar, medium concentration is required. Modified guar gum used in printing gives good printing, finishing, evenness in printing, penetration of colour, as well as it gives good colour yield and brightness.

Carpet printing is done by various methods and the gum requirement varies with the method. All of these methods require a gum thickener, and the thickener must have special characteristics for optimum performance. The thickener should be low in insoluble content, high in viscosity, easily dispersible to a lump-free paste, extremely soluble for easy washout, produce good apparent colour value, and be compatible with auxiliary chemicals and dyes being used.

In Paper Industry

For many years, this industry has been the main user of guar gum. For economic reasons, it has been replaced by modified starches and cellulose gums. The major use in paper making is at the wet end of the process.

The gum is added to the pulp suspension before the sheet is formed on either a Fourdriner machine (or endless wire) or a cylinder machine. The pulping process, which is designed to remove lignin and produces a fibrous cellulose pulp, also removes the large parts of hemicelluloses normally present in the wood. Guar replaces or supplements the natural hemicellulose needed in paper binding. The bursting and tensile strengths of paper are increased with the addition of guar gum. The gum also increases fold, pick and pulp hydration.

Guar gum is also used as an additive in size press or calendar operations in many paper industries. Guar gum increases the viscosity of the paste which is made from starch or, modified starch and controls the pick up. There is a chance of gelation with naked starch, but addition of guar prevents gelation.

In Mining Industry

In mining industry, guar gum can be used as depressant and auxiliary reagent, flotation agent, as foam stabiliser and also as filter. In the froth flotation process for concentrating ores, frothing and collector reagents are used to produce froth, which collects in the flotation cell.

Guar gum is also used as an settling agent. Normally large tanks or thickners are used to effect settling. By adding guar gum to the thickener, faster settling of the suspended solids can be achieved. This reduces the amount of thickener required to settle a given amount of pulp solids.

Guar gum has a good application as a filter aid in mining. Filters are required to remove suspended slimes or clay particles. But these small particles have a tendency to form a very compact filter cake, which traps water within itself and plugs the filter surface. The addition of guar gum to a pulp gives settling of suspended particles and these flocs no longer have a tendency to blind the filter screen and at the same time, they allow a quicker channelling of liquid through the filter cake.

Oil Well Drilling

Like many other water soluble gums, guar is used widely in the production of crude oil. In soil and gas production, drilling fluids are employed in various processes for fracturing and drilling operations. Addition of guar gum/modified guar gum gives many advantages (Water loss control, viscosity control, frictions reduction, lubrication and drilling of the drill bit, suspension aids and mobility control to carry out the well bore cuttings and chips). Drilling fluids and spacers exploit the viscosifier property of guar gum of suspend weighing materials. Guar gum with its high molecular weight and nearly straight chain configuration reduces the friction pressure in turbulent flow.

In Explosive Industry

Guar is a water proofing agent for ammonium or sodium nitrate stick explosives. The gum is used as gelling agent when reticulated in association with cationic metal with slurry explosives. Guar also absorbs the free water which is produced during reaction.

Industrial Water Treatment

Guar is mainly used as a coagulant aid in water treatment system after the addition of ferric sulphate and calcium hydroxide. Inorganic salts are added to remove the dissolved silica from water and guar gum increases the flocs resulting in filtration efficiency. Guar gum being an accepted food additive, can also be used for drinking water treatment plants.

In Tobacco Industry

Guar is used as a binder for fragmented tobacco fines in the manufacture of reconstituted tobacco sheet. Tobacco fines are moistured, guar gum and a humectant are passed between roller, presses as a film of appropriate thickness which after drying is stripped in continuous sheets.

Fire Fighting

Water for fighting fires can be handled faster and at a lower energy cost if friction reducing such as guar gum are added. In addition to providing a greater flow or less pressure drop in a fire fighting line, the jet from a nozzle tends to be more resistant to break-up by the wind and to concentrate the stream on a small area, all of which are important factors in fire fighting.

Agriculture

Guar is used as an hydrophilic colloid in soil treatment for reclaiming and improving saline and alkaline soils. Guar is also used in agricultural sprays. It stabilizes dithiocarbonate slurries used as an agricultural fungicide and as an accelerator and also used as a binding agent in insecticide compositions.

EXPORT OF GUAR GUM

As far the exports are concerned, export of guar gum has increased by three folds in last 10 years. India exported guar gum to the extent of 46581 tonnes in 1985-86 which increased to 102728.5 tonnes in 1997-98 (Table 1). Most of the exports were destined to Western countries of USA, German, Japan etc.The production and export figures of guar gum powder indicated maximum extent during 1995-96 (Table 1).

Table : 1. Production of guar gum powder and export figure from 1985-86 to 1995-96.

YEAR	PRODUCTION (MT)	EXPORT (MT)
1985-86	157740	46581
1986-87	96260	31990
1987-88	152460	52882
1989-90	300300	n.a.
1991-92	255090	61389
1993-94	244860	74360
1994-95	262680	67333
1995-96	318780	82948

n.a. not available

Market of guar in different fields are remarkable in India, for instance, 1995-96 data for Guar markets in India is given below:

Textile	12,000 MT
Explosive	1,500 MT
Paper	2,100 MT
Food & miscellaneous	2,700 MT
Total	18,300 MT

CONSUMPTION PATTERN OF GUAR GUM IN 1995-96 THROUGH OUT THE WORLD

(1) Food Industry	70750 Tonnes 30%
(2) Oil and Petroleum Industry	47166 Tonnes 20%
(3) Textile Industry	42450 Tonnes 18%
(4) Paper Industry	28300 Tonnes 12%
(5) Pharmaceutical Industry	23583 Tonnes 10%
(6) Others Industry	23583 Tonnes 10%
Total	235832 Tonnes 100%

MARKET TRENDS OF GUAR GUM

Based on production and export, the apparent consumption of guar gum can be found out, the trend in growth rate (per cent per annum) is 7-10%. Food Industry shares 30-40% as thickening, binding, and stabilising agent. Petroleum and Mining Industry shares around 20-25% of total consumption.

As far future demand is concerned, all the end user industries are growing up at a moderate rate. Assuming the rate of guar industry increase with the increase of end users, approx. 4,000,00 tonnes of guar gum would be required by 2004-2005, that indicates a very good market potential. Export demand is also likely to increase to about 1,13,702 tonnes in 2004-2005.

GUAR GUM AS VALUE ADDED PRODUCT

In the very beginning of guar gum manufacturing, when superior quality of guar gum was produced with indigenous technology, India exported mainly refined guar gum splits, which had very near price to guar seed. As a result, Indian industry had a very little scope to convert money into the profit. Gradually Indian industry started manufacturing guar gum powder from refined splits. There are many fields untapped where Indian industries can exploit and use guar gum product or derivatives, such as in tobacco making, coal mining, fire fighting, ceremics and to make water like (very transparent) cationic guar gum to be used in personal care product.

Hydroxy propyl and cationic guar gum projects can be successful in India, but the problem is of scarcity of raw material. Though, availability of propylene oxide, main raw material for hydroxypropyl guar gum, poses no problem in India, but price is very compared to export price. For example, price of propylene oxide in India is around Rs. 100/- per kg, whereas in foreign market, it is hardly Rs. 20/- kg. Thus, making hydroxy propyl derivative, is a question of project feasibility.

PATENTS IN GUAR GUM

Approximately 1500 patents on guar gum are registered through out the world, which are from U.S., U.K., Europe and India. These patents cover the various work on guar gum from manufacture to applications. Details on some important patents covering the following fields are given:

(1) Guar in oil well drilling,
-U.S. and German patents.

(2) Guar in paper industry,
-U.S., British and French patents

(3) Guar in explosive
-U.S., British and Canadian patents

(4) Guar in textile
-U.S. and British patents

(5) Guar in mining,
-U.S., German Japan patents

(6) Guar in building,
-U.S. and German patents

(7) Guar in cosmetic and pharmaceutical applications
-U.S. and German patents

(8) Guar in agriculture,
-U.S., Indian and German Patents

(9) Çarboxymethyl guar gum,
-U.S. and German patents

(10) Hydroxy; alkyl guar gum,
-U.S., German and Canadian patents

(11) Cationic guar gum,
-U.S. patent

ROLE OF INDUSTRIES AND MARKET POTENTIALITY

Guar gum came into being in India in 1956. In the early stage of guar gum industry, only refined guar gum splits was exported, where the husk was removed and the endosperm of the seeds was exported. Later on, came pulverised and treated guar gum, and today India is exporting some modified gums which are useful for food, paper, explosives, mining, textile and many others industries. Indian manufacturers should keep the quality in their products, otherwise they will be out from export market. All end users now specify their requirements and they are fully aware of the quality of the products.

Introduction of raw tamarind kernel powder (TKP) and modified TKP in the textile printing, guar gum business in textile industries were reduced too much, due to the high prices of raw guar seeds/splits during 1997-98 and 1998-99. In the year 2000-2001, though guar price would come down; the textile market can not be replaced by the guar product again, as the cost of TKP product is still less compared to guar products.

The biggest consumption of modified guar gum in textile printing is reduced now. In U.S. and European markets, it is hydroxypropyl guar gum, however the use of hydroxypropyl guar gum has also come down. Hence, export market having become unremunerative, the manufacturers have started local marketing, which also appears saturated now. In this juncture, manufacturers have started cutting each other and nobody was even trying to cover the cost. Many small industries have closed down, which were profit earner during 1981 in Punjab, Haryana, Rajasthan, Gujarat and Mumbai.

There is another important aspect of guar industries, which is the competition with Pakistan. For several reasons, Pakistan has advantages over India as far export is concerned. In Pakistan 3-4 big guar industries are controlling market, whereas, in India, every year new guar gum industries are commissioned and equal number or even more guar gum units are closed down.

Farmers are getting price for the guar seed, and the price of raw material is difficult to be controlled. The middle man brokers are playing good to make money from farmers as well as from the industries, as a result guar gum industries are now in ill shaped. Further, development of Indian Industries are restricted due to lack of In-house application research centre. The industry people do not put many resources for this purpose. Their primary aim is to produce only commodity products as per specification of the buyer in foreign countries. The in-house research centre is necessary to find out the following aspects:

(1) Improved in quality and economically feasible products,

(2) new applications of the products and

(3) value added products.

For this, continuous efforts are required in in-house research centre. Further more, there is lack of interaction between research institutes and the industries. Funds can be made available through APEDA for such type of joint programmes.

SUMMARY

Guar is an important industrial crop for India and Pakistan as well, and is considered an important foreign exchange earner because guar gum, its derivatives and high protein seed meal are sold world over. Export of guar gum swelled to 102728.5 tonnes during 1997-98, amounting to Rs. 5459.9 million. Guar gum was generally consumed in food (70750 t), oil and petroleum (47166 t), textile (42450 t), paper (28300 t), pharmaceutical (23583 t) and other industries (23583 t), statistically 30, 20, 18, 12, 10 and 10%, respectively, during 1995-96. Approximately 1500 patents on guar gum have been registered through out the world mostly in US, UK and India. Introduction of raw tamarind kernel powder (TKP) and modified TKP in textile printing, and high price of row guar seeds/splits gave guar gum business in textile industries a big blow during 1997-98 and 1998-99. The guar seed price is likely to reduce considerably during 2001-2002, still textile market can not be replaced by guar product as the cost of TKP product would still be less compared to guar product.

CHAPTER - 12

SEED PRODUCTION

S. Lal

1. IMPORTANCE OF SEED

The importance of quality seed in crop production has been recognized since the civilization and cultivation of crops had started. The seed is one of the prime inputs, which raises the productivity and quality of produce in crops. The quality seed of improved varieties also ensures other agro-inputs to be productive and most effective. In fact, this is the only input, which gives maximum return per rupee investment. This has been amply demonstrated not only through the introduction and development of dwarf varieties of wheat and rice; hybrids of maize, sorghum, sunflower, cotton, pearl millet and castor, but improved varieties of arid legumes also. The use of quality seed also increased the productivity to the extent of 15-20%. Therefore, there is no substitute of this input in the crop production programme.

The importance of quality seed further increases in the crops which are orphan and are cultivated by the small and marginal farmers in fragile environment where they cannot afford to use the costly inputs because of their economic status on the one hand and risky environmental conditions on the other. Pulses in general and arid legumes, in a particular, belong to this category of crops.

2. QUALITY SEED

As stated above, the importance of quality seed was recognized as early as the human civilization started. The people used to select the best plant/ear/panicle for next year's planting. In earlier days only the yield potential and good-looking seeds used to be the sole criteria of quality seed. As the science of seed i.e., seed technology grew the parameters of quality seeds were developed. Accordingly, the quality of seed is assessed by such parameters, which include physical and genetic purity, viability, germination, vigour and seed health. Since the germination does not reflect field performance potential of a seed lot under adverse environmental conditions, the seed vigour as a quality parameter has gained significance, particularly in grain legumes.

Seed vigour is a complex character and for its estimation a number of tests/parameters have been developed over the period. However, a practical seed vigour test should provide reproducibility of results, which could be easily interpreted and gives a good indication of field performance potential. Another very important component of seed quality is genetic purity, which is generally tested through grow out test. A number of released varieties of arid legumes have a narrow genetic base, thus, making varietal identification a difficult task for quality seed. Seed maturity, time of harvesting and the method of threshing are the other factors which determine the quality of the arid legumes.

3. PRODUCTION NETWORK

Before the green revolution era, the seed production in the country was not well organized. Only few public (National Seeds Corporation, State Farms Corporation of India and Departments of Agriculture in different states) and private seed producing organizations were engaged in the seed production. Vegetables, ornamental and horticultural plants and flowers used to be given priority. The seed production was not based on recognized norms and procedures and without following the certification. The production used to be based on without the indent from the user community. During eighties when the Indian Council of Agricultural Research (ICAR) launched National Seeds Programme (NSP) through World Bank support the systematic seed production programme was started. Sufficient infrastructure for seed production was created through this support in the ICAR Institutes, State Agricultural Universities, N.S.C., SFCI, State Seed Corporations, State Seed Certification Agencies, Seed Production Farms of State Department of Agriculture and for private seed companies. Infrastructure facilities created included development of foundation seed production farm, creation of irrigation facilities, construction of cold storage rooms for buffer stocking the seed, provision of seed processing equipment and machinery, seed planters, harvesters and threshers, equipping of seed testing laboratories. The breeder seed production (BSPs) units and Seed Technology Research (STR) Units were created at SAUs and ICAR Institutes where core staff was provided. With these facilities the seed production network has been so strengthened, that India has become largest seed production network in the world, so much so, it is helping neighbouring countries in providing quality seed of important crop varieties suitable to them.

Almost all the major states in the country have established State Seed Corporations for production and marketing of seeds of improved varieties and State Seed Certification Agencies for controlling the quality of the seed through certification. Besides the public seed producing organizations, the growth of the private seed companies has also been phenomenal. But, unfortunately, the private seed companies are involved mostly on hybrid seed production and they are least involved in the seed production of self-pollinated crops including arid legumes. Even the public seed producing organizations attach last priority for seed production of arid legumes. This resulted in poor distribution of improved varieties among the farmers and the rate of seed replacement has been far below the recommended norm.

4. CATEGORIES OF SEEDS

In the seed production chain the following categories are involved.

4.1 Nucleus Seed

This is the first category of seed in the seed multiplication chain, which is produced by the original breeder/Institution. The procedure of nucleus seed production is that initially 400-500 plants true to the type of the concerned variety are selected from the nucleus seed production plant. These plants are separately harvested and threshed. The seed of any plant differing in phenotypic traits from the parental variety is rejected. The remaining plants are raised in single plant rows. Again, if any, plant progeny is showing variation/ segregation, the whole progeny is removed. From remaining plant progenies 400-500 plants are selected for next year's nucleus seed production. Thus, all the left over progenies are harvested in bulk and threshed. The seed obtained from this bulk is called nucleus seed, which is the source for raising the breeder seed.

4.2 Breeder Seed

This is the progeny of nucleus seed produced by the original/Institution or sponsored breeder/Institution. The responsibility of monitoring of this seed production has been assigned to ICAR, which involves its institutes, State Agricultural Universities (SAUs) and other organizations engaged in research work on arid legumes. The seed production plots of this category of seed is monitored by the committee comprising of representative of State Seed Certification Agency, representative of the concerned crop Coordinator and member of the Institute producing the seed. The breeder seed is the source of foundation seed production.

4.3 Foundation Seed

This seed is produced from breeder seed by the State Seeds Corporation, National Seeds Corporation, State Farms Corporation of India, State Agricultural Universities and other public and private farms. The seed production plots are duly certified by the concern State Seed Certification Agency. The bulk of the seed is also produced by the above corporations through progressive farmers. This seed is the source of production of certified seeds,

4.4 Certified Seed

This seed is supplied to the farmers for raising the commercial crop and is produced from dul certified foundation seed mostly at the farmers' fields under the supervision of State Seeds Corporations. Similar to foundation and, this is also certified by the State Seed Corporation Agency.

5. TECHNOLOGY OF QUALITY SEED PRODUCTION OF ARID LEGUMES

Not much attention has been paid in the past for the quality seed production of pulses and arid legumes. However, the technology has been developed and perfected for quality seed production of arid legumes and the same has been given below:

5.1 Selection of Areas

Dry areas with assued irrigation facilities are considered ideal places for quality seed production for most of the crops. In addition, for arid legumes the areas with normal soil are essential as these crops are sensitive to soil salinity and alkalinity. It is nice that arid legumes are cultivated in dry areas and therefore, the areas where these crops are adopted should be used for seed production. Crop-wise such areas are listed below:

(1) **Cluster bean:** Mahendragarh, Bhiwani and Sirsa region of Haryana; Churu, Bikaner, Sikar, Jodhpur, Jaipur, Ganganagar regions of Rajasthan and Western regions of Gujarat.

(2) **Horse gram**: Drier parts of Orissa and Karnataka.

(3) **Moth bean**: Entire drier region of Rajasthan and Gujarat.

(4) **Cowpea**: Entire drier parts of Rajasthan and Karnataka.

5.2 Soil

The soils of the fields where seed multiplication of these crops has to be carried out should be normal with irrigation facilities. The area should be well protected from the stray cattle. The field conditions essential for seed certification should be met out.

5.3 Isolation distance

In order to prevent seed admixture and the chances of cross pollination, adequate distance between the fields of varieties of the same crop should be left unsown or be used for raising other crops. Since, all the arid legumes are self-pollinated 5 and 3 m isolation distance is given for raising foundation and certified seeds, respectively.

5.4 Source of the Seed

For raising breeder, foundation and certified seeds, duly certified seed of nucleus, breeder and foundation categories by the recognized agency, respectively from the seed producing organizations should be procured.

5.5 Certification

The breeder seed production is monitored by the Indian Council of Agricultural Research, for this, a committee is duly constituted by the Project Coordinator (Arid Legumes). The foundation and certified seeds are certified by the State Seed Corporation Agencies. After certification and processing, the tags for the respective seed categories, are issued by the Seed Certificating Agency.

5.6 Storage

The nucleus and breeder seeds after processing and packaging are stored by the seed producing organizations in the cold stores, mostly provided to them under Material Seeds Project/ Seeds Division of Govt. of India. The Seeds Division, thereafter, takes the responsibility of disposal of the seeds to indenting organization on payment basis.

The foundation and certified seed is stored by NSC, SFCI, State Seed Corporations and SAUs in the stores specially designed for this purpose. These categories of seeds are marketed with uniform price throughout the country. The price of breeder seed of different crops is fixed by the ICAR every year. The price of foundation and certified seeds is fixed by the NSC in consultation ICAR, SAUs and State Seed Corporations.

6. ROLE OF DIFFERENT AGENCIES IN SEED PRODUCTION

It is to mention here that India has the largest network for quality seed production. Different agencies and organizations have been assigned responsibilities for production of different categories of seeds, the same have been given below:

6.1 ICAR

Indian Council of Agricultural Research is responsible for nucleus and breeder seed production and seed technology research in SAUs and its institutes. In each crop workshop the concerned Coordinator allot the breeder seed production to concerned scientist/Institute as per variety-wise indent received from the Seeds Division of Government of India, which initially gets the indents from different organizations for production of foundation seeds.

6.2 Government of India

For monitoring the production and distribution of different categories of seeds, Government of India has created a fulfledged division (Seeds Division). This division collects variety-wise indent for breeder seed from the State Departments of Agriculture, NSC, SFCI and State Seed Corporations and the pooled indents are passed on to ICAR for production. It also monitors the production and distribution of foundation and certified seeds.

6.3 State Agricultural Universities

For the production of foundation seed, enough facilities have been created at SAUs through World Bank supported (NSP) project. For example, foundation seed production farmers have been developed in terms of leveling, fencing, irrigation facilities, processing plant and seed stores. As a commercial venture, they produce foundation seed for State Seed Corporation.

6.4 National Seeds Corporation

This is a premier seed producing organization in the country. It takes up the foundation and certified seed production at its farms and also at progressive farmer's fields. Initially it is used to involve in the production of quality seeds of vegetables and flowers. Now it has extended its seed production activities to hybrids and improved varieties of self-pollinated crops also.

6.4 State Farms Corporation of India

It is a large public seed producing organization in the country. It has 12 large farms distributed throughout the country. It is involved in the production of test stock, breeder, foundation and certified seeds of cereals, millets, pulses and oilseeds. It is also involved in the variety popularization for which it produced seeds for minikits. The major quality seeds of pulses, arid legumes and cotton is being produced by this organization.

6.5 State Seed Corporations

At state level, this corporation produces the foundation and certified seeds on its farms and through progressive farmers. The biggest seed corporation in the country is U.P. Seed and Tarai Development Corporation.

6.6 State Seed Certification Agency

This agency is involved in the quality control of the seeds through certification. The foundation and certified seed production fields are registered with this agency for certification on area basis. The seed certification inspectors of this agency regularly visit the fields and also monitor the processing of the seed and issuing the tags.

7. SEED DEMAND

In order to replace the all varieties by the newly developed improved ones, the rate of seed replacement in self-pollinated crops has been contemplated 10%. This norm holds true for arid legumes and most of the pulses. Compared to cereals, millets and oilseeds the rate of seed replacement has been very low in arid legumes. This is because of lack of both awareness and availability of sufficient quantity of quality seed of these crops, although a number of improved varieties have been developed and released for cultivation.

At the contemplated level of 10% seed replacement rate, the targeted quantities of different categories of seed has been worked out for the same, are presented in Table 1.

Table 1: Targeted Quantity of Different Categories of Seeds of clusterbean for the Next Five Years

Year	Area (Million ha.)	Breeder Seed (q)	Foundation Seed (q)	Certified Seed (q)
2000	2.40	40.0	1200	36000
2001	2.64	44.0	1320	39600
2002	2.70	48.3	1450	43500
2003	2.97	49.5	1485	44550
2004	3.27	54.5	1635	49050

Seed rate @15 kg/ha^{-1}, seed multiplication rate 1:30 and seed requirement worked out on the basis of 10 % replacement rate and 10% increase in every year.

In clusterbean the required quality of breeder seed is being produced but this seed is not fully utilized in the production of foundation and certified seeds. This results in poor spread of the improved varieties. Therefore, there should be strict monitoring of the seed production at each stage.

SUMMARY

Quality seed have great importance in the crops grown in fragile agro-ecosystems on marginal and sub-marginal lands with less inputs. Seed quality is reflected by the purity, vigour and germinability for which many tests are available. A very good network on quality production and its distribution is working in India in almost all the crops and different regions. Nucleus and Breeder seed production done under the direct supervision of the concerned breeder, requires utmost care in respect of good quality soil, adequate number of irrigations and the agronomic inputs. However, problem in indenting of seeds and lifting by the State Agencies are there which hamper and disrupt production chains. It has been estimated that in case of clusterbean almost 48.3, 1450 and 43500 q Breeder, Foundation and certified seeds would be required during 2002, whereas by 2005 the same figure would swell to 54.5, 1635 and 49050 q to cater 10% replacement rate of the old seeds and 10% increase in the area every year. Breeder seed is, however, being adequately produced but problem appears to be in final production of certified seeds.